ENGINEER'S TOOLKIT

A FIRST COURSE IN ENGINEERING

MATLAB® 5.0 for Engineers

Joe King

AN ADDISON-WESLEY SELECT™ EDITION

MATLAB for Engineers

Joe King
Department of Electrical Engineering
University of the Pacific, Stockton, California

MATLAB for Engineers

An imprint of Addison Wesley Longman, Inc.

Menlo Park, California · Reading, Massachusetts · Harlow, England
Berkeley, California · Don Mills, Ontario · Sydney · Bonn · Amsterdam · Tokyo · Mexico City

Senior Acquisitions Editor: T. Michael Slaughter
Developmental Editor: Judy Ziajka
Editorial Assistant: Royden Tonomura
Production: Caroline E. Jumper
Copyeditor: Robert Fiske
Proofreader: Holly McLean Aldis
Indexer: Nancy Kopper
Cover Design: Yvo Riezebos
Text Design: Side by Side Studios
Composition: Fog Press

This is a module in the Engineer's Toolkit, an Addison-Wesley SELECT edition. Contact your sales representative for more information.

Photo Credits:
Chapter 1: NASA/LBJ Space Center
Chapter 2: © J. Richardson/Westlight
Chapter 3: © David Parker/Science Photo Library/ Photo Researchers, Inc.
Chapter 4: © R. Ian Lloyd/Westlight

Library of Congress catalog card number 95-131339.

ISBN: 0-201-35094-7

1 2 3 4 5 6 7 8 9 10—CRK—01 00 99 98 97

Addison Wesley Longman, Inc.
2725 Sand Hill Road
Menlo Park, CA 94025
http://www.aw.com/cseng/toolkit/

Contents

1 Introducing MATLAB 5

Space Travel

Today, the 1969 Apollo moon landing seems like science fiction when you consider the relatively primitive computer technology available three decades ago. Still, using the limited circuit integration techniques available, Apollo's designers managed to build an onboard computer measuring only 6 by 12 by 24 inches, roughly the size of today's typical PC.

The computer that guided the astronauts to the moon and back used only 36,864 15-bit words of read-only program memory and 2,048 15-bit words of read/write memory, a total of about 72 kilobytes. Today, a typical PC contains several megabytes of read-only memory and 8 to 64 megabytes of read/write memory. In addition, most modern PCs include a hard disk that stores at least another 1,000 megabytes. In the application at the end of this chapter, you will see how MATLAB could have been used to help design the Apollo computer memory.

INTRODUCTION

MATLAB is a technical computing language for high-performance numeric computation and data visualization. MATLAB combines matrix computation, graphics, numerical analysis, and signal processing in an easy-to-use environment where the user solves complex problems without the overhead of traditional programming languages. MATLAB can be used to solve problems in any field that involves mathematics, including engineering, physics, chemistry, applied mathematics, and finance. Within engineering, application areas include signal processing, system identification, and control systems. The emphasis in MATLAB is on visualization, the most effective approach to solving many complex problems. MATLAB's two-, three-, and four-dimensional (animated) graphing capabilities give you, the engineer, a powerful problem-solving tool.

The name *MATLAB* stands for "matrix laboratory." The basic data element used in MATLAB is a *matrix*. This matrix structure enables MATLAB users to write programs that solve computational problems in a small fraction of the time that it would take to write equivalent programs using computer languages such as C or C++. Although the syntax of MATLAB is C-like, particularly when creating loops and performing I/O, you should find that learning MATLAB is much easier and more enjoyable than learning C. MATLAB programs tend to be short and readable, and can be written using any text editor or word processor, as well as the intelligent built-in MATLAB Editor/Debugger application. (For beginning users, the Editor/Debugger is the most important improvement of MATLAB 5 over MATLAB 4.)

The main goal of this MATLAB module is that you become competent at solving and graphing the kinds of mathematical problems that you will likely encounter as a lower-division engineering student. The problems covered here range from simple arithmetic to calculus.

Two secondary goals of this module are (1) that you learn to enjoy using MATLAB, so you will *want* to use it to solve your problems and (2) that you know the range of MATLAB's capabilities, even if you cannot yet use its advanced features. In this way, you may be motivated to go beyond this module and learn how to use MATLAB's advanced features in your future work as an engineering student and practicing engineer.

This module describes MATLAB 5 for Windows 3.x and Windows 95. Versions are also available for Macintosh and UNIX systems.

This chapter introduces you to problem solving by first examining the types of problems MATLAB can solve. It presents a five-step problem-solving process that is useful for solving typical engineering problems, and then uses a simple example to illustrate the use of the five-step process. The chapter ends with an overview of the conventions and features of MATLAB.

1-1 PROBLEM SOLVING WITH MATLAB

MATLAB is one of the most easily learned yet most powerful engineering tools you could have at your disposal. Simple arithmetic is easy enough. Just enter an expression or equation at the MATLAB command prompt, ».

For example, to perform simple arithmetic, you enter an expression or equation at the MATLAB command prompt as follows:

```
» 5*6
```

MATLAB computes the product and returns the result, which, unless you create your own variable, is stored in the default variable *ans*.

```
ans=
  30
```

To create your own variables, you simply type them. For example, if you enter

```
» product=5*6
```

MATLAB responds

```
product=
  30
```

Chapter 2 introduces you to the basics of MATLAB. It explains how to begin and end a MATLAB session, how to access window and menu options, and how to use the Help features. It describes the ease with which you can create and use matrices, the basic building blocks of MATLAB. The chapter then discusses the creation of 2-D graphs. Finally, Chapter 2 explains how to create interactive user programs that run within MATLAB.

Chapter 3 covers engineering computations. It describes some of MATLAB's more commonly used built-in functions, array operators, and matrix operators. The chapter also explains how to write your own functions and how to use loops in your programs. The chapter emphasizes learning to write MATLAB programs and functions.

In Chapter 4, you will learn, among other things, how to solve simultaneous equations, create random numbers, and perform statistical analysis. For example, the following commands determine the mean of the integers 1 through 5:

```
» numbers=[1   2   3   4   5];
» mean(numbers)

ans=
  3
```

Chapter 4 also covers how to determine derivatives and integrals. The chapter closes with a section on plot creation. MATLAB can easily produce plots such as the following, which took two commands to create:

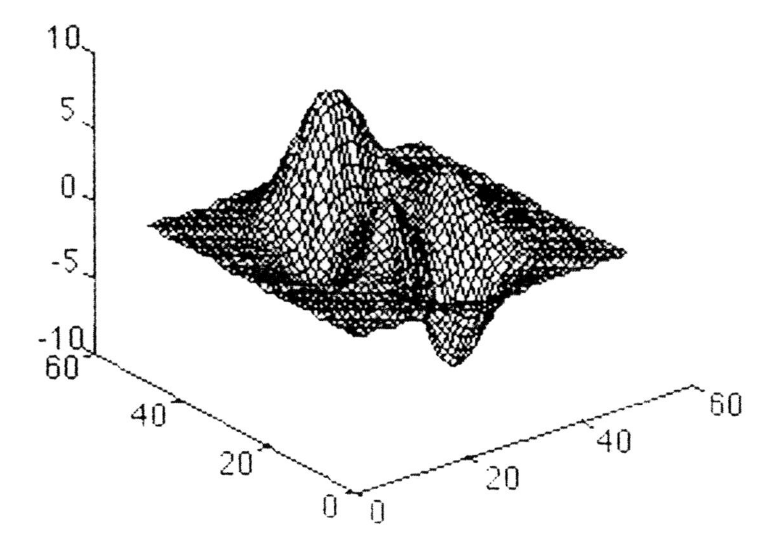

MATLAB can plot single functions, multiple functions, polar plots, 3-D surface plots, and even 4-D plots (the fourth dimension is time) that use animation and sound. MATLAB will automatically scale axes for you, or you can choose your own scales. Graph axes can be linear or logarithmic. You can add titles, axis labels, and text to your graphs, which will then print exactly as they appear on the screen. The result is a nice, neat record of your work that you can proudly insert into any engineering report. Furthermore, since MATLAB is a Windows application, you can easily copy your work into any Windows-based word processing program, such as Microsoft Word for Windows or WordPerfect for Windows.

1-2 THE FIVE-STEP PROBLEM-SOLVING PROCESS

Problem solving is what engineering is all about. For centuries, engineers have applied scientific principles to the solution of real-world problems. Thousands of years ago, Egyptian engineers solved some of the problems involved in building the pyramids. Contemporary engineers have addressed the problems of spanning large rivers with bridges and establishing worldwide communication systems.

However, it seems that for every problem engineers solve, society asks them to solve two more. Fortunately, to this end, computers have significantly increased the efficiency of engineers. Computers never forget what they are taught, never get bored or tired, and never complain about working 24 hours a day. They can collect data and, using programs such as MATLAB, analyze and graph that data.

Whether or not an engineer uses a computer, he or she must approach each problem methodically using an *algorithm*. An algorithm is a step-by-step procedure by which one arrives at a solution to a problem. For example, a cooking recipe is an algorithm.

Typically, when an engineer uses a computer, the algorithm must be described to the computer using a programming language such as C or C++. MATLAB requires little programming; it contains powerful built-in algorithms that users can call up via simple commands. Users can write

their own programs to supplement those provided by MATLAB. In any event, all MATLAB programs, whether built in or created by the user, tend to be much more concise than equivalent C or C++ programs, providing extensive data analysis and graphics capabilities in relatively few lines of code.

Whatever the problem, a general design process involves the following five steps, which have been adapted for solving engineering problems with MATLAB:

General Problem-Solving Procedure	MATLAB Problem-Solving Procedure
1. Define the problem clearly.	Define the problem.
2. Gather information.	Refine the problem definition.
3. Generate and evaluate potential solutions.	Research potential solutions.
4. Refine and implement a solution.	Choose and implement a solution.
5. Verify the solution through testing.	Test the solution.

1-3 APPLYING THE FIVE-STEP PROBLEM-SOLVING PROCESS

Step 1: Define the Problem. For the designers of the Apollo spacecraft onboard guidance computer, defining the problem was a bit difficult because the computer was being designed in parallel with the spacecraft and the rest of its contents. Initially, all they knew was that the computer had to be as small and light as possible. It had to include both read/write memory and read-only memory. The read/write memory would store sensor data and temporary results of computations. The read-only memory would store the guidance program, which was expected to require roughly one-half million bits of storage.

Step 2: Refine the Problem Definition. As the Apollo project progressed, the engineers were told that the onboard computer must fit into a space of roughly one-half cubic foot and weigh less than 75 pounds. The engineers determined that, for accurate control, the sensor data would require 15 bits per word and that they must store approximately 2,000 words of data values. The programmers found that the control program would require about 37,000 words of storage if each word consisted of 15 bits.

Step 3: Research Potential Solutions. The engineers could have implemented the computer processor entirely with transistors, or even vacuum tubes; however, both take up more space and consume more power than did available integrated circuits (ICs). They would have liked to have used nothing but ICs; however, many of the ICs they required had not yet been developed. The computer memory, too, could have been implemented using transistors; however, core memory was used. (In a

core memory, each bit is stored in a tiny donut made of magnetic material.) Integrated circuit memories had not yet been developed, whereas core memories were common. Core memories were much smaller than were transistor memories and consumed far less power.

Step 4: Choose and Implement a Solution. To maximize compactness and minimize power consumption, the engineers decided to implement the computer processor using available simple ICs and transistors. They decided to keep the user interface as simple as possible to save space and weight. For input, the computer used a touch pad; for output, it used four one-line digital displays.

Ultimately, the engineers determined that the available space for the entire guidance computer was 6 by 12 by 24 inches. The computer had to store 2,048 15-bit words of data in read/write memory. A core memory cell (bit) size was determined to be 8.6×10^{-4} in^3. Assuming the processor size was 6 by 12 by 17 inches, we now use MATLAB to compute the amount of program memory the Apollo engineers can give the programmers:

```
» mem_cell_size=8.6e-4;
» total_space=6*12*24;
» proc_space=6*12*17;
» total_memory_space=total_space-proc_space;
» total_memory=total_memory_space/mem_cell_size

total_memory=
5.8605e+005

» data_memory=2048*15;
» program_memory=(total_memory-data_memory)/15

program_memory=
  3.7022e+004
```

From the calculations, there was room within the available 6 by 12 by 24 inches for 37,022 words of program memory. The Apollo engineers actually implemented 36,864 15-bit words. In order to safeguard the program, they decided to make the program memory read-only. The memory had to be hand-wired, laboriously creating the program, bit by bit.

The Apollo programmers divided the guidance program into three main parts: the attitude state vector, the reaction control system laws, and the thrust vector control laws. This approach was an example of *top-down design*, a process that involves examining the problem as a whole and then breaking it down into smaller problems. The process continues until it results in problems that can be readily defined and solved. In this way, a large task can be completed more quickly, by several designers, working in parallel. Designing separate, independent modules also aids in debugging both hardware and software.

Step 5: Test the Solution. The Apollo engineers installed the guidance computer in a prototype Apollo spacecraft and then tested it. The engineers thus ensured that the program and data would fit into the available memory. However, it was impossible for them to test the computer as adequately as they would have liked. And if anything went wrong out in space, there would be no radio downloading of software patches since the program memory was not writable. Although there were guidance problems in Apollos 11, 13, and 14, the guidance computer worked amazingly well throughout the entire lunar space program.

As a future engineer, you must learn to make the five-step problem-solving process second nature in solving engineering problems. Each of the remaining chapters of this module offers an application of this five-step process. The discipline and domain of each application problem are as follows:

Applications Across the Disciplines

Discipline	Application Problem	Chapter
Electrical engineering	Optical fiber propagation delay	2
Computer engineering	CPU instruction set design	3
Electrical engineering	Electronic circuit mesh analysis	4

1-4 MATLAB FEATURES

MATLAB is a very powerful software package. Listed here are some of the features and capabilities of MATLAB 5 for Windows. With the limited space we have in this module, we will be able to cover only some of these features. In any event, most of MATLAB's capabilities are more appropriate for upper-division and graduate-level engineering students, as well as practicing engineers.

User Interface

- Pull-down and pop-up menus that aid beginning and advanced users
- Mathematical rules checking on all user entries
- Error messages that flag specific parts of any errant user entry
- Graphs that can be copied into other Windows documents
- Context-sensitive online help

Computational Capabilities

- General equation solving, including simultaneous equations
- Trigonometric, hyperbolic, exponential, Bessel, and many other elementary functions
- Complex numbers, complex variables, and complex functions
- Summation and product operations

- Matrix operations, including multiplication, inverses, transposes, and determinants
- Statistical functions, including mean, standard deviation, linear regression, gamma function, and erf
- Polynomial analysis
- Numerical integration and differentiation
- Solution of ordinary differential equations
- Matrix decomposition and factorization
- Signal processing functions for frequency domain analysis, filter analysis, and filter design
- Control system functions for system modeling, design, and analysis
- Interpolation and curve fitting, including spline curve fitting
- Random number generation
- Program control flow instructions, including the *if* statement, *for* loops, and *while* loops

Graphing Functions

- 2-D contour, histogram, line, mesh, polar, polygon, and scatter plots
- 3-D line, mesh, polygon, scatter, and wireframe plots
- Shaded surfaces, lighting, and Cartesians objects
- Animation and sound
- Multiple traces per graph
- Axis, line, and text manipulation

Programming Language Features

- C-like programming commands
- C code generation (with MATLAB compiler)
- Control structures (if-then-else, for, while)
- Program debugging
- String manipulation
- ASCII and binary file input/output
- Reading and writing of workspaces
- User-defined functions

1-5 MATLAB TOOLBOXES

As you will discover in Chapter 3, you can easily create your own MATLAB functions (called M-files) to supplement those that are built into the package. In addition, you can purchase from MathWorks, the makers of MATLAB, a variety of optional toolboxes, sets of useful functions (M-files) that MATLAB programmers and others have created over the years. These application-specific sets of MATLAB functions cover a variety of disciplines, including the following:

- The *Signal Processing Toolbox* adds to MATLAB's basic set of functions commands for one-dimensional and two-dimensional digital signal processing. It also includes functions for the design and analysis of digital filters and for performing Fourier analysis.

- The *Optimization Toolbox* contains commands for the optimization of general linear and nonlinear functions and for the solution of nonlinear equations.
- The *Control System Toolbox* includes commands pertaining to control and systems engineering.
- The *Robust-Control Toolbox* provides commands for robust control system design.
- The μ-*Analysis and Synthesis Toolbox* includes functions for the design and analysis of robust linear control systems using μ-analysis and synthesis techniques.
- The *System Identification Toolbox* adds commands for parametric modeling and system identification.
- The *Spline Toolbox* includes commands for using splines to model functions of an arbitrary nature. Splines are useful for generating curves that best provide a function for a set of data points.
- The *Neural Networks Toolbox* adds functions for designing and simulating neural networks.

These toolboxes give MATLAB what is probably its most important feature: its extensibility. As more toolboxes become available, often from scientists and engineers, they are added to MATLAB's repertoire and made available to present and future MATLAB users.

1-6 SIMULINK

In addition to toolboxes, MathWorks offers an optional program called SIMULINK. MathWorks calls SIMULINK, "The ultimate graphical modeling, simulation, and prototyping environment." SIMULINK is an extension of MATLAB that provides modeling, analysis, and simulation of physical and mathematical systems. This graphical, mouse-driven program enables users to create and interactively manipulate designs in block diagram form on the computer screen. You simply "drag and drop" components from an extensive block library to build models of dynamic systems quickly and easily. You can run simulations of your designs, displaying results "live" using scope and graph blocks. Using MathWork's Real-Time Workshop, you can generate C code from SIMULINK models for embedded applications and rapid prototyping of control systems.

1-7 NEW MATLAB 5 FEATURES

MATLAB 5 includes a number of new features that, compared to MATLAB 4, make it a more complete, self-contained development and analysis environment. Among these new features are

- An integrated visual editor/debugger (MATLAB 4 uses any external text editor)
- An online help desk and documentation system that works via a web browser (provided by the user) and the TCP/IP protocol

- Additional data types and data structures (including a structure data type)
- More than 200 additional math and data analysis functions
- Additional 3-D plot types (including 3-D bar graphs and 3-D pie graphs)
- Faster, more accurate graphics capabilities
- Programming enhancements (including the addition of a *case* statement)

An inexpensive, limited student edition of MATLAB 5 is available. For more information on the student edition or standard edition of MATLAB 5, its toolboxes, or SIMULINK, visit the MathWorks web site at http://www.mathworks.com. You can also contact MathWorks at: The MathWorks, Inc., 24 Prime Park Way, Natick, MA 01760-1500. Phone: (508) 653-1415. Fax: (508) 653-2997. E-mail: info@mathworks.com.

SUMMARY

This chapter introduced you to MATLAB by briefly describing its problem-solving capabilities and listing its features and options. The chapter presented a five-step problem-solving process and an application of the process. The five steps are as follows:

1. Define the problem.
2. Refine the problem definition.
3. Research potential solutions.
4. Choose and implement a solution.
5. Test the solution.

Key Words

algorithm
matrix
top-down design

2 Getting Started with MATLAB 5

Networking

Computer networking is the interconnecting of computers for the purpose of sharing information, computing power, and peripherals. Peripherals include hard disks, monitors, printers, and plotters. These peripherals and the computers they serve are called network nodes.

Most computer networks are confined to a few offices, a single building, or a localized set of buildings. These smaller networks are called local area networks (LANs). Larger networks, called metropolitan area networks (MANs), may be dispersed over an area the size of a city. Still larger networks, called wide area networks (WANs), may be distributed throughout a nation or the world.

LAN nodes are typically personal computers (PCs), whereas MAN and WAN nodes are often minicomputers and mainframes, as well as PCs. LAN nodes are usually interconnected by twisted pairs of wire or coaxial cables called "links." MAN and WAN links are typically telephone lines, optical fiber cables, or microwave communication channels. We explore the implementation of an optical fiber-based WAN in the application problem at the end of this chapter.

INTRODUCTION

In this chapter, you will learn the basics of MATLAB. The first section discusses the conventions used in this module, how to start and end a MATLAB session, and how to save your work. The section ends with an overview of the MATLAB command window and a description of the Help feature.

Section 2-2 begins with an explanation of how to use scalars, expressions, and variables in MATLAB. It then covers the declaration and manipulation of matrices and vectors. Next, it explains how to access information stored in the *workspace*, which is the memory space where MATLAB stores the names and values of all variables. An introduction to X-Y graphing concludes the section.

Section 2-3 introduces programming and file handling. The section begins with an explanation of how to get input from the user and how to send unformatted and formatted output to the monitor screen. Then it explains how to write MATLAB programs and how to read and write data files, concluding with a discussion of printing hardcopy outputs. All the subjects covered in Sections 2-2 and 2-3 are covered in more detail in the remaining chapters of this module.

You will find that you use MATLAB to solve engineering problems in a way that is similar to the way you might express them on paper. Therefore, solving problems using MATLAB is generally much faster than solving those problems using a computer language such as C or C++. As you work your way through each example and Try It! exercise in this module, you may be pleasantly surprised at how quickly you learn to use MATLAB and how much fun it is to use.

2-1 GETTING STARTED

Module Conventions

This module uses the following conventions:

1. CAPITAL LETTERS represent filenames and directories.
2. *Italics* represent variables, functions, commands, and key words.
3. The `typewriter font` represents characters you are instructed to type, that is, MATLAB commands. MATLAB responses are also shown in `typewriter font`.
4. All user entries are indicated by preceding them with the » prompt. It is assumed that you press the ENTER key at the end of each entry.
5. **Bold** print is used to denote vectors and matrices.
6. When told to "select OPTION," you should click the left mouse button on the named option. For example, if you were told to "select Open under the File menu," you would click the left mouse button on "File" on the menu bar near the top of the command window and then on "Open" in the File menu.
7. MATLAB treats all numbers and sets of numbers, including scalars and vectors, as matrices. (A *scalar* is treated as a matrix with a single row and column; a *vector* is viewed as a matrix with a single row or column.) However, to avoid confusion, in this module, the term *matrix* will refer only to matrices that contain multiple rows and columns.

You can perform nearly all MATLAB operations using *shortcut key operation*. For example, you can quit MATLAB by *sequentially* pressing the keys ALT,f,x. Pressing the ALT key activates the pull-down menus, pressing *f* activates the File menu, and pressing *x* causes MATLAB to exit to Windows. Some command shortcuts use the CONTROL key, which is pressed *simultaneously* with another key. For example, CONTROLc copies selected text to the clipboard, CONTROLv pastes clipboard text to the MATLAB command line, and CONTROLq quits MATLAB. MATLAB indicates shortcut keys by underlining them in the names of the menus and the selections under the menus, or by listing them (as in "Ctrl+X") to the right of a menu entry.

Generally, this module will instruct you to use the mouse to perform MATLAB operations since "select Exit MATLAB under the File menu" is more descriptive than "press ALT,f,x."

Running MATLAB

To run MATLAB 5 for Windows 3.x or Windows 95, you must first run Windows. Chances are, Windows (3.x or 95) came up automatically when you turned on your computer. If not, type the command

```
win ENTER
```

at the DOS prompt.

Once you are up and running in Windows, either a MATLAB icon, labeled "MATLAB 5," or a MATLAB Program Group icon (a square) should be present.

If a MATLAB Program Group icon is present, you must double-click the mouse on it to gain access to the MATLAB icon. Once you see a MATLAB icon, double-click on it to run MATLAB. The MATLAB command window (described later) will appear and within it, the MATLAB prompt », where you begin typing your MATLAB commands. You should run MATLAB now and familiarize yourself with it as it is described throughout the remainder of this chapter.

When MATLAB first starts, its window may be somewhat smaller than full screen. If you desire a larger size, click the mouse on the appropriate button in the extreme upper right-hand corner of the window. The MATLAB window will expand to full screen.

If, upon running MATLAB 5, the error message, "Unable to initialize MIPC," appears, you will have to install the Internet protocol, TCP/IP, on your PC. If you enter the command *help mipc*, MATLAB will tell you how to do this. However, the task is best left to a Windows expert. MATLAB 5 uses TCP/IP to perform a number of tasks, including the creation of its Help feature user interface. This interface is based on a user-provided web browser. Therefore, your PC must also have a web browser installed on it for you to gain the full benefit of the MATLAB Help feature.

If you choose to experiment with MATLAB at this point, note that if you get an unexpected, lengthy response from MATLAB, you can press CONTROLc to terminate that response. Pressing CONTROLc aborts any MATLAB command, function, or program.

Quitting MATLAB and Saving Your Work

There are a number of ways you can quit MATLAB.

1. » quit
2. » exit
3. » CONTROL q
4. » ALT ,f,x
5. Select Exit MATLAB under the File menu.
6. In Windows 3.x, double-click on the small box in the upper left-hand corner of the MATLAB window. In Windows 95, click on the x in the upper right-hand corner of the command window.

Before you quit, you may want to save your work using the *save* command. MATLAB will save the contents of the workspace, that is, the names and values of your variables, in the default directory in a file called MATLAB.MAT. When you restart MATLAB, you can start where you left off using the *load* command. MATLAB will read MATLAB.MAT, restoring all variable declarations and values. For example,

```
» save
```

saves the entire workspace in the file MATLAB.MAT, whereas

```
» load
```

reloads the workspace at the start of your next session.

You do not have to use the default directory, and you do not have to use the default filename, MATLAB.MAT. You can save the workspace on any drive, in any directory, with any legal filename, by specifying the desired drive, path, and filename. For example,

```
» save a:mywork
```

saves the workspace on a floppy disk in drive A: in the file MYWORK.MAT, whereas

```
» load a:mywork
```

reloads the workspace from that floppy disk. Note that DOS filenames are limited to eight characters.

Another option is saving the workspace in a specified directory on the hard disk. For example,

```
» save c:\class\matlab\jones
```

will save the workspace in the directory C:\CLASS\MATLAB in the file JONES.MAT. You can also save your workspace by selecting Save Workspace under the File menu and restore your workspace by selecting Load Workspace under the File menu.

Whenever you begin work on a new problem, you should always clear the workspace by using the *clear* command. Otherwise the workspace, which includes all previously declared variables, will remain in effect and can alter the results of the new problem. For example, if you declared a variable $X = 2$ in a previous problem, and you forget to declare a new variable $X = 6$ when starting a new problem, MATLAB will calculate your results based on the value of X declared in the previous problem.

The *clear* command will remove all variables from the workspace. To remove selected variables while leaving others in place, just name the variables you want to remove after entering the *clear* command. For example, if you want to remove only the matrix **X** from the workspace, type

```
» clear X
```

If you want to retain the variable names and values in the workspace, but clear up the clutter on your screen, use the *clc* or *home* command, which is equivalent to selecting Clear Session under the Edit menu. The *clc* and *home* commands simply move the cursor to the upper left-hand corner of the command window, giving you a cleared command window; they do not affect the workspace.

The MATLAB Command Window

The following figure displays the MATLAB window.

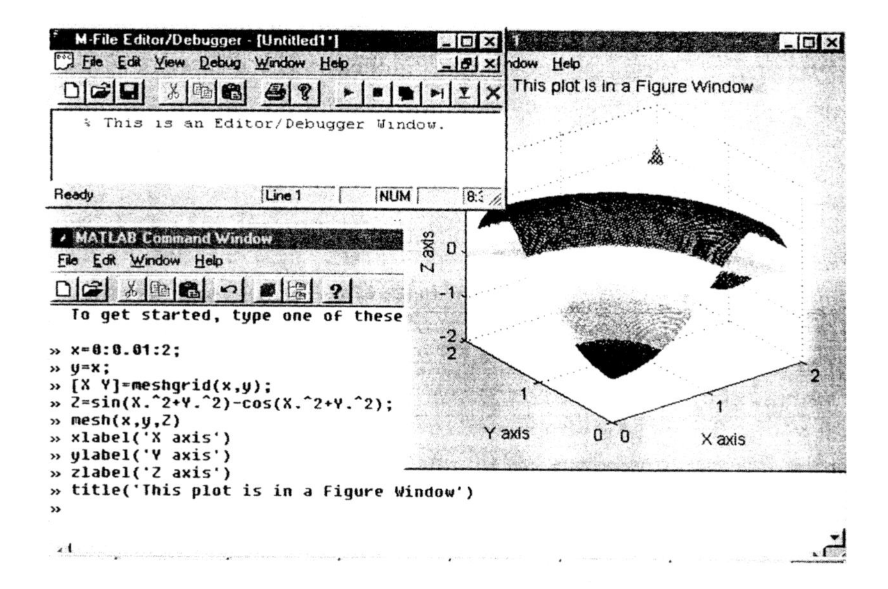

As this figure shows, MATLAB uses three windows: the command window, the figure window, and the editor/debugger window. The *command window* accepts and processes MATLAB commands, one at a time (similar to the way you interact with DOS on a PC). The *figure window* displays graphical plots. The *editor/debugger window* is used to create MATLAB programs and functions. Most interaction with MATLAB takes place within the command window. When you first run MATLAB, the command window is the only window shown. The figure window appears only when you create a plot. The editor/debugger window appears only when you edit a MATLAB file.

The top line of the MATLAB command window, called the *title bar,* is titled, "MATLAB Command Window." The second line at the top of the command window, the *menu bar,* includes four pull-down menus: File, Edit, Windows, and Help. (Earlier versions of MATLAB include an Options menu, which has been replaced by the Preferences option under the File menu.)

The File menu includes the following selections:

New	Creates a new MATLAB program or function
Open...	Opens an existing program or function for editing (same as the *edit* <filename> command)
Open ""	Opens for editing an existing program or function, the name of which is selected in the command window
Run Script...	Executes the commands listed in a text file (similar to executing a DOS batch file)
Load Workspace...	Restores a workspace that has been stored in a file named by the user (similar to the *load* command)
Save Workspace...	Saves the workspace in a file named by the user (similar to the *save* command)
Show Workspace...	Shows the contents of the workspace, that is, the names and values of all variables (similar to the *whos* command)
Set Path...	Establishes default directory paths
Preferences...	Sets default command window numeric formats and text fonts
Print Setup...	Initializes the printer
Print...	Prints the contents of the command window
Printer Selection	Prints a selected portion of the command window contents
Exit MATLAB	Quits MATLAB

The Edit menu includes the following selections:

Undo	A standard undo command, not often useful
Cut	Cuts selected text to the clipboard
Copy	Copies selected text to the clipboard
Paste	Pastes text from the clipboard
Clear	Clears selected variables from the workspace
Select All	Selects the entire contents of the command window
Clear Session	Clears the command window, without affecting the workspace contents (using the *clear* command in the command window has the opposite effect: the contents of the workspace are cleared without affecting the command window contents)

The Cut, Copy, and Paste menu options are standard Microsoft Windows editing features. Like the Undo option, they are not particularly useful in MATLAB. You can copy previous commands that still appear in the command window by dragging the mouse across them and then selecting Copy. But a typically easier way to access previous command entries is to search through the command buffer in which MATLAB stores all commands entered during a session. You can work your way back and forth

through the buffer by pressing ↑ and ↓, respectively. This allows you to recall a previous entry and edit it, eliminating the tedious task of retyping lengthy commands. Pressing the ESC key clears the command line so you can enter a new command. Pressing CONTROL a moves the cursor to the beginning of a command line, whereas pressing CONTROL e moves the cursor to the end of the line.

The last selection in the Edit menu, Clear Session, clears the command window. But an easier way to clear the window is to use the *clc* or *home* command (discussed at the end of this subsection). Most MATLAB users find that they rarely use the Edit menu. However, one might use it to select the entire contents of the command window and then cut and paste those contents into a word processor document as a record of how a particular task was accomplished.

The Window menu lists the windows that are presently open. The default window is the MATLAB command window. When a graphing command is executed, either at the MATLAB prompt or from within an executing program, a figure window is created to contain the graph. Each retrieved MATLAB program, function, or data file resides in an editor/debugger window. Clicking on the name of a window takes you to that window.

Below the menu bar is the tool bar. The *tool bar* allows you to quickly perform popular functions by merely clicking on a button. From left to right, the tool bar buttons are New M-file, for creating a new MATLAB program or function; Open M-file, for editing an existing MATLAB program or function; Cut; Copy; Paste; Undo; Workspace Browser, for checking the contents of the workspace; Path Browser, for setting default directory paths; and ? for opening the Help Window.

Getting Help

You can access the MATLAB Help feature at any time during a MATLAB session. Under the Help menu are the selections Help Window, Help Tips, Help Desk (HTML), Examples and Demos, About MATLAB, and Subscribe (HTML). Selecting Help Window gives you a directory of MATLAB commands, from which you can get help on how to use MATLAB commands. For example, selecting Help Window under the Help menu, and then selecting matlab/general, and then selecting help, will give you the same results as will entering *help help* at the command line. Selecting Help Tips under the Help menu offers very few tips.

The Help Desk option offers the bulk of MATLAB's help assistance; however, this assistance is available only if you have an installed web browser, such as Netscape's Navigator or Microsoft's Explorer. (Explorer seems to be MATLAB's preferred browser.) A typical MATLAB installation includes many megabytes of help, stored in HTML (hypertext markup language) files. HTML files are best read using a web browser. Also note that your PC must have the TCP/IP Internet protocol installed, even if the PC is not attached to a computer network.

The Examples and Demos option under the Help menu gives you the opportunity to work your way through many excellent example problems. These problems range from the very simple to the complex. The demos

display a number of beautiful graphical MATLAB results. The About MATLAB option gives version and other information about the MATLAB version you are using. The Subscribe option uses your web browser to help you subscribe to the MathWorks newsletter.

To remove any of the Help windows, click on the x (Windows 95) in the upper right-hand corner of the window.

Whenever you run MATLAB, it begins by displaying the following message:

```
To get started, type one of these commands: helpwin, helpdesk,
or demo
```

The *helpwin* command is equivalent to selecting Help Window under the Help menu. The *helpdesk* option is equivalent to selecting Help Desk under the Help menu. The *demo* command is equivalent to selecting Examples and Demos under the Help menu.

For example, as a beginner, you should type

```
» demo
```

to run MATLAB Demos, a program that offers example problem solutions and demonstrations of MATLAB's advanced capabilities. The *intro* command will take you directly to some tutorial material. The Gallery option will display some very nice graphics. There are even several MATLAB-based games that you can play.

Typing

```
» help help
```

displays an explanation of the Help feature.

Typing

```
» whatsnew matlab
```

provides a description of MATLAB 5's many new features.

If your computer is connected to the Internet, the Help Desk provides a connection to The MathWorks MATLAB web site. At the web site, you can use electronic mail to ask questions, make suggestions, and report possible bugs. You can also use the site's Solution Search Engine to query an up-to-date database of technical support information.

2-2 INTRODUCTION TO MATLAB

Expressions and Variables

An *expression* (for example, $x + 4/5$) is a mathematical construct that has a value or set of values. Like most programming languages, MATLAB uses expressions; however, unlike other programming languages, these expressions always involve entire matrices.

The value of an expression can be found in MATLAB simply by typing the expression and pressing ENTER. The following involves two 1×1 matrices:

```
» 4/5
```

MATLAB responds

```
ans=
  .8
```

If you wish, you can assign your own variable name to an expression (for example, $a = 4/5$). If you do not specify a variable name, MATLAB will store the result in the default variable, *ans*, as shown. MATLAB variable names consist of a letter, followed by any number of letters, digits, or underscores. MATLAB, however, uses only the first 31 characters of a variable name.

If an expression contains variables, the value of the expression can be computed only if the values of the variables have been previously given. For example, if you want MATLAB to calculate the value of $y = x + 5$, you must first supply a value for x, as follows:

```
» x=2

x=
  2

» y=x+5

y=
  7
```

If you have not supplied a value for x, MATLAB will return an error message.

```
» y=x+5

??? Undefined function or variable 'x'.
```

MATLAB uses conventional decimal notation, with an optional decimal point and leading plus or minus sign. *Scientific notation* uses the letter *e* to specify a power-of-ten scale factor. Imaginary numbers use either an *i* or *j* as a suffix. Some examples of legal MATLAB numbers are

```
25
-16
3.7e-28
-0.9e41
1i
2j
```

Note that MATLAB is case sensitive when using variables and function names. For example, x and X are seen as two different variables. MATLAB's built-in functions use all lowercase characters in their names; the command *sqrt*(10) returns the square root of 10, but *SQRT*(10) and *Sqrt*(10) will return error messages.

Another built-in function that we will use in this introductory section is *exp*. The function *exp*(x) returns e^x. Two other built-in functions that we will use in this section are *sin*(x) and *cos*(x), which return the sine and

cosine, respectively, of *x*. Many of MATLAB's most commonly used built-in functions are described in Section 3-1. To get a list and brief description of MATLAB's most commonly used functions, enter the command *help elfun*. Some of MATLAB's functions are *built in*; others are implemented in M-files, as subprograms, which use the built-in functions to implement more complex functions. An *M-file* is an ASCII text file that contains MATLAB code that implements a MATLAB program or function.

The precedence of arithmetic operations within an expression is the same in MATLAB as in most other programming languages. The order of precedence is as follows:

1. Parentheses
2. Exponentiation, left to right
3. Multiplication and division, left to right
4. Addition and subtraction, left to right

When using MATLAB, you must keep the default precedence of arithmetic operations in mind. The simple problem of computing the average of two numbers demonstrates the problem.

```
» age1=27;
» age2=71;
» average=age1+age2/2

average=
  62.5000

» average=(age1+age2)/2

average=
  49
```

Clearly, the second answer is the correct one. If addition had precedence over division, both calculations would have resulted in the answer 49.

When using MATLAB, you must pay close attention to the use of semicolons. A semicolon at the end of a MATLAB command suppresses the listing of the results of the execution of the command. Note that in the example, in which we computed the average of two ages, the semicolon at the end of each of the first two lines prevented MATLAB from echoing the value of age1 and age2. Omitting the semicolon would have resulted in

```
» age1=27

age1=
  27

» age2=71

age2=
  71
```

It's not uncommon for a user to forget to add the semicolon before pressing (ENTER). This is not a problem in a simple exercise such as the one shown. However, it's not uncommon for users to forget to add a semicolon after creating a large matrix or inputting a large data file and then have MATLAB start listing thousands of element values. If this should happen to you, you can use the (CONTROL)c command to stop the listing.

Semicolons can also be used to separate multiple commands on the same line. For example, the three commands used in the preceding problem could have been entered on a single line, as follows:

```
» age1=27; age2=71; average=(age1+age2)/2
```

You can also use commas to separate multiple commands on a single line. However, the comma does not suppress a response as does the semicolon.

EXAMPLE 2-1

Evaluating an Expression

Find the values of the following expression:

$$16e^{-5t} \quad \text{for} \quad t = 5$$

SOLUTION

As you solve the example problems of this chapter, remember to press (ENTER) at the end of each command line and to use the *clear* command as needed to clear out all past variables before starting each new problem. Make sure you end the following command with a semicolon to tell MATLAB not to echo the results of the command.

```
» t=5;
```

In the next command, you do *not* use a semicolon.

```
» 16*exp(-5*t)
```

MATLAB responds with

```
ans=
  2.2221e-010
```

MATLAB automatically displays very small and very large numbers in exponential format. The result was 2.2221×10^{-10}.

As we noted earlier, the two commands used in this example could have been entered on one line.

```
» t=5; 16*exp(-5*t)

ans=
  2.2221e-010
```

Try It

Declare the following variables, assigning the indicated values:

$$a = 1.12$$

$$b = 2.34$$

$$c = 0.72$$

$$d = .81$$

$$v = 19.83$$

Use the preceding variables to determine the value of the following expressions:

$$1 + b/v + c/v^2$$

$$(b - a)/(d - c)$$

$$1/(1/a + 1/b + 1/c + 1/d)$$

Matrices

As stated earlier, the basic element in MATLAB is the matrix. That is, MATLAB works only with matrices, handling even scalars and vectors as matrices.

A matrix can be explicitly declared as a list of elements contained within brackets ([ ]), with elements separated by spaces or commas, and rows separated by semicolons. For example, a 2×4 matrix can be declared by typing

```
» M=[1 2 3 4; 5 6 7 8]

M=
   1 2 3 4
   5 6 7 8
```

You could have obtained the same result by entering

```
» M=[1,2,3,4; 5,6,7,8]
```

Matrices can also be created by calling up MATLAB programs (implemented in M-files) or by reading data files. (Programs and data files are discussed in Section 2-3.) Finally, matrices can be created via matrix operations and functions (covered in Section 3-1).

Individual matrix elements are specified with indices within parentheses, as in M(1,1). The first number in parentheses specifies the row; the second specifies the column. Row and column indices begin at 1; therefore, M(1,1) is the element in the upper left-hand corner of the matrix.

```
» M(1,1)

ans=
   1
```

If **M** is an $m \times n$ matrix, then referencing M(i,j), where i is less than 1 or greater than m, or j is less than 1 or greater than n, yields an error message. For example, if **M** is a 2×4 matrix, M(4,4) does not exist, and the following occurs:

```
» M(4,4)

???  Index exceeds matrix dimensions.
```

In formal mathematical notation, matrix names are usually designated by an uppercase letter, whereas individual matrix elements are designated by lowercase letters. However, since MATLAB is case sensitive, you must consistently use the same case whether designating a complete matrix or just one of its elements. The following MATLAB session demonstrates the principle:

```
» M=[0 1 2 3; 4 5 6 7; 8 9 0 1]

M=

  0 1 2 3
  4 5 6 7
  8 9 0 1

» M(2,3)

ans=
  6

» m(2,3)

??? Undefined function or variable 'm'.
```

Matrix elements can be expressions, even expressions that use functions. For example, a 2×2 matrix can be declared as follows:

```
» A=[4/5   sqrt(2);   exp(-1)   cos(pi/8)]

A=
  0.8000 1.4142
  0.3679 0.9239
```

Try It Create the following matrix, and then determine the sum of each row, column, and diagonal:

8	1	6
3	5	7
4	9	2

You can create matrices from vectors, a process called *concatenation.* For example, if **a**, **b**, and **c** are all row vectors of the same length, they can be combined to form a three-row matrix **D** as follows:

```
» a=[1 2 3]; b=[4 5 6]; c=[7 8 9];
» D=[a; b; c]

D=
  1 2 3
  4 5 6
  7 8 9
```

Note that the vector names are separated by semicolons. As when creating any matrix, semicolons delimit rows; therefore, **a** is the first row, **b** is the second row, and **c** is the third row.

The following is another example of concatenation:

```
» x=[1 2 3 4];
» y=x+1;
» z=[x;y]

z =
    1.00    2.00    3.00    4.00
    2.00    3.00    4.00    5.00
```

Here, the matrix **x** became the first row of the matrix **z**, and the matrix **y** became the second row of the matrix **z**.

By deleting rows or columns from a matrix, you can create smaller matrices from larger ones. An empty pair of brackets ([]) is used to delete a matrix row or column. The following example creates a 3 × 3 matrix **x**, sets the matrix **y** equal to **x**, and then deletes the middle column of the matrix **y**. A colon is used to designate either all rows or all columns. In the following example, y(:,2) refers to all rows of column 2 of the matrix **y**.

```
» x=[1 2 3; 3 4 5; 4 5 6]

x =
    1.00    2.00    3.00
    3.00    4.00    5.00
    4.00    5.00    6.00

» y=x;
» y(:,2)=[]

y =
    1.00    3.00
    3.00    5.00
    4.00    6.00
```

Matrices of text elements can also be created in MATLAB. The following creates a 3×4 matrix of character elements:

```
» c=['Jill'; 'Mary'; 'Sam ']

c=

  Jill
  Mary
  Sam
```

Note that, as with all matrices, the rows must all be the same size; therefore, a space was added at the end of Sam so that each row included four elements, each a single character.

EXAMPLE 2-2

Creating Text Matrices Using Concatenation

Create a matrix that contains the names James, Dare, and Allison. Create a second matrix that contains the words first, second, and third. Remember to pad shorter terms with blanks so that the rows of each matrix all have the same number of character elements. Concatenate the two matrices so that the names form a left column and the positions form a right column. Display the result.

SOLUTION

```
» names=['James   ';'Dare    ';'Allison '];
» position=['first ';'second';'third '];
» result=[names position]

result=

  James   first
  Dare    second
  Allison third

» size(result)

ans=

  3    14
```

Note that the matrix **result** is a 3×14 matrix, not a 3×2 matrix. Also note that here you combined two text matrices to form a larger text matrix; however, you cannot combine text and numerical matrices.

Try It

Create a matrix that contains the oil reservoir names West Side, Lower Kern, Lost Hills. Create a second matrix that contains the well numbers 11-1, 13-2, and 14-1. Concatenate the two matrices so that the reservoir names form a left column and the well numbers form a right column. Display the result.

Vectors

Vectors can be created in exactly the same way as are matrices. The following examples, respectively, create a row matrix and a column matrix:

```
» R=[1 2 3 4]

R=

   1   2   3   4

» C=[1; 2; 3; 4]

C=

   1
   2
   3
   4
```

As with matrices, the rows are separated by semicolons.

Vectors can also be created by specifying beginning and ending values separated by a colon. For example, the following command creates a vector that ranges from 1 to 5:

```
» P=[3:7]

P=

   3   4   5   6   7

» P(1)

ans=

   3
```

You can enter the vector as P=3:7, excluding the brackets. The default increment is 1. If a different increment is desired, it must be specified between the beginning and ending values, as shown next. In the first of the following two examples, the increment is -2; in the second example, it is $\pi/4$.

```
» P=10:-2:4

P=

   10  8   6   4

» Q=pi:pi/4:2*pi

Q=

   3.1416  3.9270  4.7124  5.4978  6.2832
```

You can use the *linspace* function to let MATLAB compute the increment for you; you just specify how many elements you want generated. The *linspace* function requires three arguments: the starting value, the ending value, and the number of desired points. The following creates a vector that contains five elements, evenly spaced between $-e^1$ and e^1:

```
» E=linspace(-exp(1),exp(1),5)

E=

   -2.7183   -1.3591   0    1.3591    2.7183
```

Try It Use the *linspace* function to create a vector of ten elements that range, equally spaced, from $-\pi/8$ to $+\pi/8$.

You can create a smaller vector by extracting it from a larger one by specifying the range of indices of the desired elements, separated by a colon, within parentheses. For example, you could extract the first three elements of a six-element vector as follows:

```
» M=[2 4 1 3 5 0];
» N=M(1:3)

N=

   2  4  1
```

You can extract vectors from matrices by deleting all but one row or column. As stated, a colon is used to designate either all rows or all columns. In the following example, A(:,3) refers to all rows of column 3 of matrix **A**, and A(1,:) refers to all columns of row 1 of matrix **A**.

```
» A=[1 2 3 4; 5 6 7 8; 2 4 6 8]

A=

   1  2  3  4
   5  6  7  8
   2  4  6  8

» B=A(:,3)

B=

   3
   7
   6

» C=A(1,:)

C=

   1  2  3  4
```

You will find the capability of extracting columns and rows from matrices useful when graphing data. For example, a data file may contain a two-column matrix, in which the first column contains measurements and the second column contains times that correspond to the measurements. In order to plot measurements versus times using MATLAB, you would plot the first column against the second after converting each to a vector.

EXAMPLE 2-3

Creating Vectors from Matrices

Create a vector from the first column of matrix **x**, naming the vector **time**. Then create a vector from the second column of matrix **x**, naming the vector **stress**.

$$\mathbf{x} = \begin{matrix} 1 & 3.2 \\ 2 & 4.4 \\ 3 & 4.8 \\ 4 & 3.8 \\ 5 & 4.1 \\ 6 & 3.9 \end{matrix}$$

SOLUTION

First, define the matrix **x** (which more likely would be read from a text data file, as shown in Section 2-3). Then create the matrices **time** and **stress**.

```
» x=[1 3.2; 2 4.4; 3 4.8; 4 3.8; 5 4.1; 6 3.9]

x =
   1.0000    3.2000
   2.0000    4.4000
   3.0000    4.8000
   4.0000    3.8000
   5.0000    4.1000
   6.0000    3.9000

» time=x(:,1)

time =
     1
     2
     3
     4
     5
     6

» stress=x(:,2)

stress =
    3.2000
    4.4000
    4.8000
    3.8000
    4.1000
    3.9000
```

The semicolons were omitted after each command in order to see the results of each command. The command *plot(time, stress)* would produce a graph of stress versus time. Creating graphs is covered later in this section.

Try It Create the matrix **x** shown in the preceding example. Then create a column vector in which each row element is the sum of the two elements in each corresponding row of **x**.

Accessing Workspace Information

During an extended interactive MATLAB session, there may be times when you want to know what variables you have declared so far. The *who* command lists all existing variables in the workspace. The following example clears the workspace, declares a few variables, and then uses *who*.

```
» clear
» x=1:9;
» t=5;
» t/2

ans =

   2.50

» y=t*[-1 -2; 2 1];
» who

Your variables are:
ans     t     x     y
```

The *whos* command gives more complete information about the variables in the workspace.

```
» whos

Name     Size     Bytes     Class
ans      1x1      8         double array
t        1x1      8         double array
x        1x9      72        double array
y        2x2      32        double array

Grand total is 15 elements using 120 bytes
```

You can see from the table that each element in a matrix uses 8 bytes of memory. Note that Size is shown in rows by columns; so, for example, **x** is a nine-element row vector. Also note that t is listed as a 1×1 matrix; as stated earlier, MATLAB views all variables, including scalars, as matrices.

If you just want to know the dimensions of a particular matrix in your workspace, use the *size* command. For example, to determine the dimensions of the matrix **x**, type

```
» size(x)

ans=
  1 9
```

The matrix **x** is a 1×9 matrix; it has one row and nine columns.

When a function returns more than one argument, and you want to assign their values to variables, you must place the variable names in a vector. For example,

```
» [numrows numcols]=size(x)

numrows=
  1
numcols=
  9
```

EXAMPLE 2-4

Accessing Workspace Information

Compare the memory required to store the square root of 2 to that required to store the square root of -2.

SOLUTION

```
» clear; clc
» real=sqrt(2); imag=sqrt(-2); whos

Name    Size    Bytes    Class
imag    1x1     16       double array (complex)
real    1x1     8        double array

Grand total is 2 elements using 24 bytes
```

Try It

Determine the amount of memory required to store the following identity matrix:

$$\begin{matrix} 1 & 0 & 0 & 0 \\ 0 & 1 & 0 & 0 \\ 0 & 0 & 1 & 0 \\ 0 & 0 & 0 & 1 \end{matrix}$$

An identity matrix is a matrix that contains 1s on the upper left to lower right diagonal and 0s elsewhere.

X-Y Graphs

Because of its impressive graphics capabilities, MATLAB excels in allowing the engineer to visualize data. X-Y, or 2-D, graphs are normally the most commonly used type of graph, and they are used throughout the remainder of this module to demonstrate concepts. Other graph types, such as polar and 3-D plots, are covered in Chapter 4.

When creating a plot (or graph), you must first decide on the resolution you desire. Low resolution results in a curve that is not as smooth as you might like; high resolution results in greater memory usage and slower graph generation. For example, the following low-resolution graph plots only six points of one cycle of *sin*(*x*):

```
» x=0:2*pi/5:2*pi;
» y=sin(x);
» plot(x,y)
```

The figure window should now show

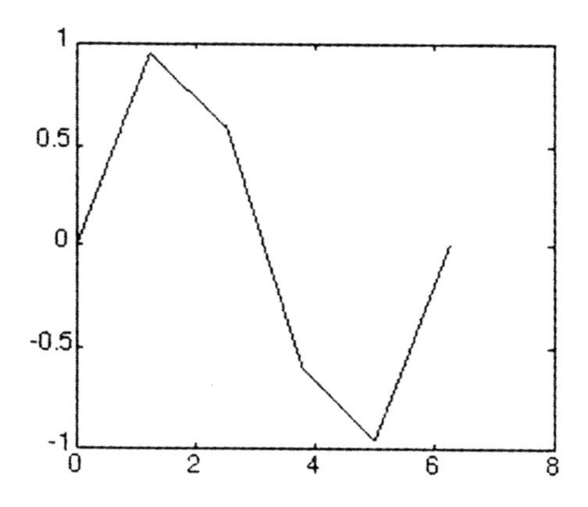

The next graph also plots one cycle of *sin*(*x*), increasing the resolution by plotting 101 points.

```
» x=0:2*pi/100:2*pi;
» y=sin(x);
» y=sin(x);
```

And we now have

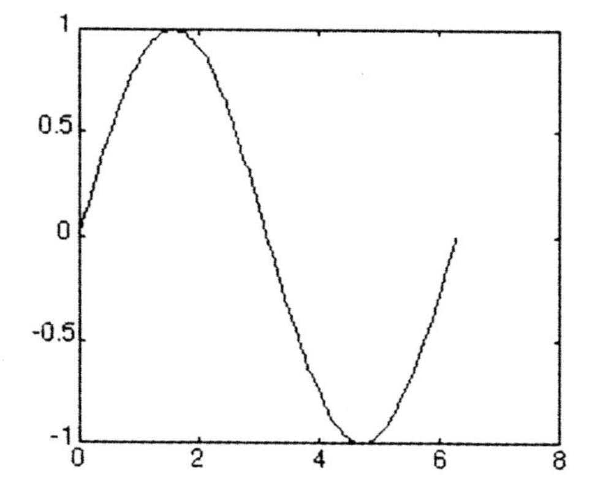

Note that you can print a hardcopy of any MATLAB plot by selecting Print under the File menu of the figure window.

In Example 2-5, you will plot one vector against another to solve an engineering economics problem. The *plot(x,y)* function is used, which produces a linear plot of the elements of vector **y** versus the elements of vector **x**.

The example introduces the *hold on* command. You can create multiple plots within a single figure window by using the MATLAB *hold on* command to freeze the present and any subsequent plots. Its counterpart, the *hold off* command, unfreezes the plots, and the next plot will replace all previous plots.

EXAMPLE 2-5

Graphing One Vector Against Another

The cost of producing a product generally decreases with the number of units produced. Of course, revenue increases with the number of units sold.

In engineering economics, the breakeven point is that number of units produced and sold when costs equal revenues. Assume that analysis has shown that the cost, in dollars, of producing widgets is given by the equation $cost(w) = 1{,}300 + 5.60 \cdot w$. That is, it costs \$1,305.60 to produce one widget, but it costs \$6.90 per widget to produce 1,000 widgets. If we sell our widgets for \$8.00 each, our revenues are defined by the equation $revenue(w) = 8 \cdot w$.

Plot the cost and revenue functions in order to determine the breakeven point.

SOLUTION

Define the range of the independent variable *widgets*, the number of widgets produced.

```
» widgets=0:1000;
```

Define the two relevant functions.

```
» costs=1300+5.60*widgets;
» revenues=8*widgets;
```

Now, plot the two curves.

```
» plot(widgets,costs)
» hold on
» plot(widgets,revenues)
```

The figure window should now show

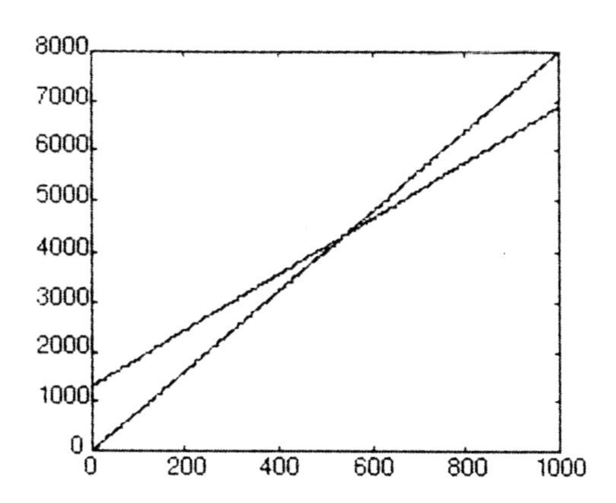

It can be seen from the preceding graph that the breakeven point occurs approximately when 500 widgets are manufactured. The costs and revenues at that point are approximately $4,000. Mathematical analysis would yield more accurate answers; however, an engineer often uses graphical analysis for quick, rough answers when accurate input figures, such as those for costs and revenues, are not available.

Note that, although all MATLAB command entry is performed in text mode, when the plot was requested, MATLAB created a figure window in which the plot appeared. Also note that *plot* automatically scaled both axes linearly. Semilog and log plots can be created using the *semilogx*(*x,y*), *semilogy*(*x,y*), and *loglog*(*x,y*) functions.

MATLAB also automatically chooses appropriate axes limits. However, you can set your own limits using the *axis* function. The *axis* function takes as a parameter a single matrix that contains four elements that set the limits of the X-axis and the limits of the Y-axis. For example, if, after creating a plot, you type

```
» axis([1 2 3 4] )
```

the X-axis will range from 1 to 2, and the Y-axis will range from 3 to 4.

A third *plot* argument, enclosed in quotes, can be passed to *plot* to designate the type and color of line drawn. For example, *plot*(*x,y,'g--'*) draws a dashed green line. MATLAB can plot various line types, plot symbols, and line colors; you designate your choice with a one- to three-character third argument.

.	point	y	yellow
o	circle	m	magenta
x	x-mark	c	cyan
+	plus	r	red
*	star	g	green
-	solid line	b	blue
:	dotted line	w	white
-.	dashdot line	k	black
--	dashed line		

The default, used when you omit the third argument, is a solid line.

The *plot()* command also allows you to specify multiple plots in one command so that the *hold on* command is not necessary. Using this approach, you would have entered

```
» plot(widgets,costs,widgets,revenues)
```

A figure window often overlays work within the command window. You can move a figure window by placing the mouse cursor in the title bar of the window, pressing the left mouse button, and then dragging the window to a new location. The default size of a figure window is often larger than necessary. In order to decrease its size, move the mouse cursor to any of the four corners of the window, press the left mouse button, and then drag the corner inward toward the center of the window. To remove a figure window, double-click on the x in the upper right-hand corner of the figure window (Windows 95).

MATLAB graphs can be enhanced using the *title('title')*, *xlabel('label')*, *ylabel('label')*, and *grid* functions. Titles and labels, enclosed in single quotes, can be whatever you choose. The *grid* function adds grid lines to the plot. All four of these functions are used in the next example.

EXAMPLE 2-6

Graphing One Column of a Matrix Against Another Column

Plot the second column against the first column of the following matrix of data. Then, on the same graph, plot the third column against the first column.

Time (Sec.)	Voltage A (mV)	Voltage B (mV)
1.0000	5.0000	5.4000
1.1000	6.0000	6.2000
1.2000	6.7000	7.0000
1.3000	7.2000	7.5000
1.4000	8.1000	8.2000
1.5000	7.4000	7.7000
1.6000	6.6000	7.0000
1.7000	5.0000	6.0000

SOLUTION

Turn *hold* off if it is still on from the preceding example. Input the data. If you must continue a lengthy MATLAB command on a second line, enter an ellipsis (...) at the end of the first line before pressing ENTER.

```
» hold off
» data=[1.0 5.0 5.4; 1.1 6.0 6.2; 1.2 6.7 7.0;...
 1.3 7.2 7.5; 1.4 8.1 8.2; 1.5 7.4 7.7; 1.6 6.6 7.0;...
 1.7 5.0 6.0];
» plot(data(:,1),data(:,2))
```

MATLAB should create the plot at this point. Now, hold the first plot, create the second plot, and add the labels and grid. When you press the *h* as you type *hold on*, MATLAB will return you to the command window.

```
» hold on
» plot(data(:,1),data(:,3),'--')
» title('A Graph of Voltage versus Time')
» xlabel('Time in Seconds')
» ylabel('Voltage in Millivolts')
» grid
```

The figure window may be partially hidden at this point. It will be entirely hidden if the command window is full-screen size. If any part of the figure window is visible, click on the visible area, and the figure window will be brought to the foreground. If it is completely hidden, select its name, "Figure No. 1," under the Windows menu. The figure window should show

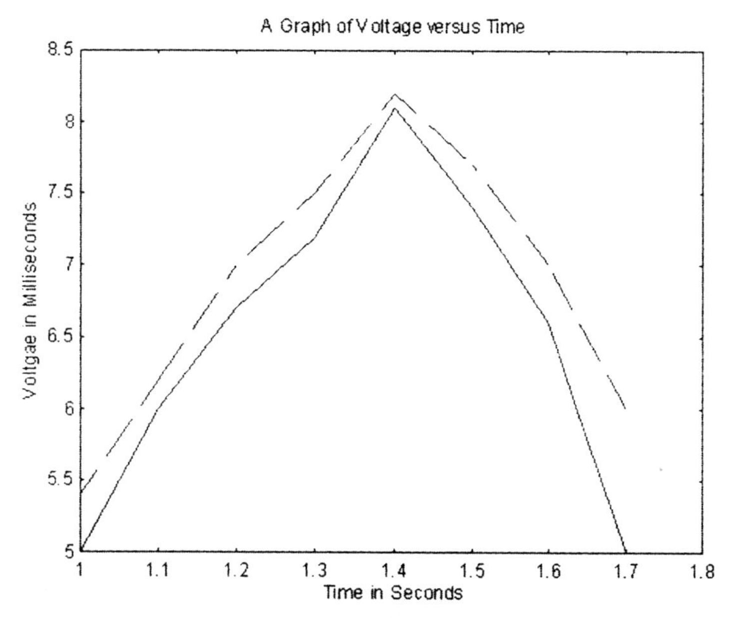

You may find dealing with figure windows frustrating at first. For example, after you've created a graph, your next key press will return you to the command window and move the figure window to the background. When you create a second *plot*, it will, by default, with *hold off* in effect, replace the previous plot in the existing figure window. However, you will not automatically see this plot because the figure window is not brought to the foreground. To bring a hidden figure window to the foreground, you must select the figure window under the Windows menu. You can resize the command and figure windows so that both can always be seen simultaneously.

Try It Plot the function $y(x) = 3\cos(x) - 4\sin(x)$.

2-3 PROGRAMS AND FILES

Input and Output

When you use MATLAB interactively to perform simple mathematical operations, you would rarely use input and output commands. However, when you become an experienced user, you will find yourself working primarily with MATLAB programs, which almost always require the use of input and output commands. Nearly all MATLAB programs require some data (input) on which to act. A program can instruct the user to input data directly, or it can read the data from data files (covered later in this section).

The *input* command is used when you wish to have users enter data directly into the program. The command requires a single text argument. For example,

```
» age=input('Please enter your age: ')

Please enter your age: 23
age=
  23
```

MATLAB programs always include at least one output command that passes the results of the program's computations to either the user or a data file. The *disp* function provides an easy way to pass results to the user when formatting is not a concern. The *disp* function accepts only a single argument, either text delimited by single quotes, a variable, or a constant. For example,

```
» age=input('Please enter your age: ');

Please enter your age: 23

» disp('Your age is '), disp(age)

Your age is
  23
```

If you try to use multiple arguments, MATLAB will return an error message.

```
» disp('Your age is ',age)

??? Error using==> disp
Too many input arguments.
```

An alternative to *disp* is *fprintf*, which formats the output. This function is taken straight from the C programming language and works the same as it does in C. Its input arguments are a format specification, enclosed in single quotes, and a list of one or more matrices, the values of which are to be printed.

In the *fprintf* format specification, %f, %e, and %g are used to indicate where in the output, and with what format, the matrix values are to be printed. You specify %f for fixed point format, %e for exponential format, and %g for either fixed point or exponential, whichever is shorter. The string \n is used to force a carriage return in the output. The following demonstrates the use of *fprintf*:

```
» celsius=0:100;

» fahrenheit=celsius*9/5+32;

» t=input('Enter a Celsius temperature: ');

Enter a Celsius temperature: 25

» fprintf('%f Celsius=%f
Fahrenheit.\n',t,fahrenheit(t))

25.000000 Celsius=75.200000 Fahrenheit.
```

Here, fahrenheit(t) is *not* a function; it merely specifies the *t*th element in the vector **fahrenheit**. Conventional programming languages normally distinguish between function calls and vector references by using, for example, parentheses in one case and brackets in the other; then, *sin*(*x*) would be a function call, whereas *sin*[*x*] would be a vector reference. MATLAB, however, uses parentheses in both cases, giving vector references precedence. Therefore, if you give a matrix the name of a built-in function, that function will no longer be available. For example, the *sin*(*x*) function would no longer be available if you create a *sin* vector as follows:

```
» sin=[1:5];
```

By default, %f and %e print to six decimal places, and %g suppresses the printing of trailing zeros. However, you do not have to use these defaults. For example, %m.nf and %m.ne print *m* digits with *n* digits to the right of the decimal point. Therefore, using %4.2f instead of %f in the preceding example would have resulted in the following output:

```
25.00 Celsius=75.20 Fahrenheit.
```

You can change the format of MATLAB output. The default, called *format short*, is fixed decimal to four decimal places. Entering the command *format long* results in output being displayed to 14 decimal places, whereas the command *format short* returns the output to the default format. Remaining options include *format short e* (exponential), *format long e*, *format hex* (hexadecimal), and *format bank* (rounded to two decimal places to indicate dollars and cents).

The *pause* command is used in MATLAB programs to temporarily suspend program execution. Note that *input* does the same thing until the user provides a value for the variable named in the *input* command. The difference is that *pause* requires no value from the user and is often used in the following way, after displaying some message:

```
» disp('Press Enter to go on...'); pause
```

The user must press (ENTER) before program execution will continue. Again, *input*, *disp*, *fprintf*, and *pause* are typically used in MATLAB programs rather than in interactive sessions. You will become more proficient in their use as you start writing programs in the next subsection.

EXAMPLE 2-7

Using Input and Output Commands

Use the *fprintf* function to print the value of π to eight decimal places, preceded by the text, "Pi to 8 decimal places is."

SOLUTION

```
» fprintf('Pi to 8 decimal places is %9.8f. \n',pi)

Pi to 8 decimal places is 3.14159265.
```

Try It

Use *fprintf* to print the value of the square root of 2 to six decimal places, preceded by the text, "The square root of 2 to 6 decimal places is."

Creating MATLAB Programs (M-Files)

As a new user of MATLAB, you will spend most of your time in interactive sessions, entering commands one at a time and receiving MATLAB responses. However, in practice, engineers more often use MATLAB to run programs they have written or purchased.

A MATLAB program consists of MATLAB commands and C-like code that are stored in an ASCII file. Program files must use the filename extension .M; for this reason, they are referred to as *M-files.*

To create a MATLAB program, it is most convenient to use the built-in MATLAB 5 Editor/Debugger; however, you can use any editor or word processor. Select the File menu, then select M-file from the New option. The editor/debugger window appears with the title, "M-File Editor/Debugger - [Untitled1]."

After you write your program in the editor/debugger window, select Save As under the window's File menu to save the program with your desired filename. As with saving the workspace, you can specify any drive and directory as the destination of the file. The default filename extension is .M, which you must use if MATLAB is going to execute your program. To exit the editor/debugger window and return to the command window, select Exit Editor/Debugger under the File menu, or double-click on the x in the upper right-hand corner of the editor/debugger window (Windows 95).

As you create and edit your MATLAB programs, note that MATLAB allows you to use the following DOS commands within the command window:

dir	List the files in the default directory
type <filename>	List the contents of the file named
delete <filename>	Delete the file named
cd	Change the default directory

EXAMPLE 2-8

Writing a MATLAB Program

Write a program that asks the user for a temperature in Fahrenheit and prints the equivalent temperature in Celsius to two decimal places.

SOLUTION

Select New under the File menu. Click on M-file in the New menu. An editor/debugger window appears. Type the following in that window:

```
% This program converts Fahrenheit temperatures to Celsius.
clear
clc
fahr=input('Please enter a temperature in Fahrenheit: ');
celsius=(fahr-32)*5/9;
fprintf('That temperature is %5.2f in Celsius.\n',celsius)
disp('Press Enter to go on...'); pause
clear
clc
```

Select Save As under the File menu, enter the filename CELSIUS.M, and press ENTER. (If you do not want to use the default drive and directory, you must also specify alternatives, such as A:CELSIUS.M.)

Select Exit Editor/Debugger under the File menu.

This closes the editor/debugger window and returns you to the MATLAB command window. Now, to run the M-file you just created, type

```
» celsius
```

You should see the following after entering 200 ENTER. If an error message appears, type *edit celcius.m* and make the necessary editions to the program.

```
Please enter a temperature in Fahrenheit: 200
That temperature is 93.33 in Celsius.
Press Enter to go on...
```

Since the program CELSIUS.M ended with the commands *clear* and *clc*, pressing ENTER after the "Press Enter to go on..." message appears will clear the screen.

You can edit M-file by selecting Open under the File menu; a directory of existing M-files will appear. Only those M-files in the default directory will be listed. You can retrieve and edit an M-file that is not listed by entering its complete path specification and name. Once you've accessed the directory, double-click on the name of the file you wish to edit. After editing, save the file.

If you are sure about the name of the M-file you want to edit, you can simply enter

```
» edit <filename>
```

at the command line.

If you want to examine a file before deciding whether you want to edit it, use the *type* command in the command window, which will list the contents of any M-file, including the comments (preceded with a %) inserted in the file. For example, any of the following:

```
» type celsius
» type celsius.m
» dbtype celsius
» dbtype celsius.m
```

lists the contents of CELSIUS.M. The *dbtype* command adds line numbers to the listed file.

Note that the *type* command will not put you into an editing mode. If you want to edit a file after examining it, you must open the file, as outlined earlier.

Entering

```
» what
```

lists all MATLAB-related files, including M-files, that are in the current default directory. You use the *which* command to locate, on your default drive, the file named. It is particularly useful for finding M-files that you have created when they are not in the default directory. Typing

```
» which fftdemo.m
```

might yield the result

```
c:\matlab\toolbox\matlab\demos\fftdemo.m
```

However, the command *which plot* will give, "plot is a built-in function."

The *keyboard* command is used as a program debugging tool. When the *keyboard* command is executed in an M-file, MATLAB ceases execution of the program commands and begins accepting user commands from the keyboard. A k» prompt appears. The user can enter any legal MATLAB command. For example, the *whos* command would indicate the state of the workspace. Typing

```
» return
```

causes MATLAB to continue execution of the M-file program. Entering a *keyboard* command in a MATLAB program is similar to entering a breakpoint in the program. A *breakpoint* is a point at which a program temporarily ceases execution to allow the user to examine or modify the values of variables.

Try It

Write a MATLAB program that asks the user to provide the coefficients a and b, in the linear equation $y = ax + b$ and then plots y in an X-Y graph in which x varies from −100 to +100. You choose the number of points plotted.

Using Data Files (MAT-Files and ASCII Files)

The data that engineers collect, analyze, and base decisions upon are often stored in data files. MATLAB can read and write two kinds of data files: MAT-files (so named because they use the filename extension .MAT) and ASCII files. ASCII files store data as ASCII characters and use the filename extension .DAT. MAT-files are preferred because they store data in memory-efficient binary form but are normally used only when MATLAB creates the data file. Programs written in other computer languages usually store data in and read data from ASCII files. Therefore, MATLAB has the ability to read and write ASCII data files so that it can pass data between its programs and programs written in other languages.

MATLAB creates a MAT-file whenever the *save* command is executed. For example, using *save* without an argument saves the workspace in the file MATLAB.MAT. The command

```
» save mydata X Y
```

saves the matrices **X** and **Y** in the file MYDATA.MAT.

To restore **X** and **Y** to the workspace in a later MATLAB session, you enter the command

```
» load mydata
```

Using the .MAT extension is not necessary. As with other MATLAB files, you can specify the drive and directory, if desired. For example,

```
» save a:\jones\mydata X Y
» load a:\jones\mydata
```

An ASCII data file must consist of a single matrix; each line in the file corresponds to a row in the matrix. Any editor or word processor can be used to create ASCII data files. The name of the file must be the same as the name of the matrix it stores. For example, to load an ASCII data file into the matrix **X**, the name of the file must be X.DAT, and you enter the command

```
» load X.dat
```

The .DAT filename extension must be used when reading or writing ASCII files; without an extension, MATLAB will assume a .MAT extension.

Remember that MATLAB is case sensitive; therefore the command

```
» load x.dat
```

will load the data in X.DAT into the matrix **x**, whereas

```
» load X.dat
```

will load the data in X.DAT into the matrix **X**.

To store the MATLAB matrix **X** in an ASCII data file, you enter

```
» save x.dat X /ascii
```

Since DOS is not case sensitive, it does not matter whether you use the name x.dat or X.dat when the *save* command is used. The command

```
» load X.dat
```

will restore the contents of **X** in a later MATLAB session.

EXAMPLE 2-9

Reading and Writing a Data File

Create a MAT-file that contains the matrix

1 2 3 4

2 3 4 5

3 4 5 6

4 5 6 7

Clear the workspace, and then load the matrix. Display the cosine of each element in the matrix.

SOLUTION

```
» clear; clc
```

Place the specified data in a matrix **X**.

```
» X=[1 2 3 4; 2 3 4 5; 3 4 5 6; 4 5 6 7];
```

Save the data in a MAT-file called MYDATA.MAT.

```
» save mydata X
```

Clear the workspace so that **X** no longer exists in memory, and clear the command window.

```
» clear; clc
```

Retrieve **X** and display the cosine of its elements.

```
» load mydata
» cos(X)

ans=
   0.5403  -0.4161  -0.9900  -0.6536
  -0.4161  -0.9900  -0.6536   0.2837
  -0.9900  -0.6536   0.2837   0.9602
  -0.6536   0.2837   0.9602   0.7539
```

After loading MYDATA.MAT, it would have been prudent to check the names and sizes of the variables it contained by entering the *whos* command.

MATLAB also includes standard C input and output functions, such as *fopen, fread, fscanf, fwrite, fseek,* and *fclose.* If you are familiar with C, you can use these routines to read and write standard binary data files that can be written and read by programs written in C or other high-level languages.

Try It

Create a vector **X** that contains ten equally spaced values from 0.1 to 1.0. Create a vector **Y** that contains the ten values e^X corresponding to the ten values in **X**. Save these two vectors in a MAT-file called MYDATA.MAT. Clear the workspace, load **X** and **Y**, and graph **Y** versus **X** in an X-Y graph. Do not delete this work; you will be asked to print the graph in the next Try It! exercise.

Printing

You can print the contents of the command window, the figure window, or the editor/debugger window by selecting Print under the File menu in the applicable window. By default, the entire window in which you select Print is printed. However, you can drag the mouse to select any portion of a command or editor/debugger window and then print the selection. Also under the File menu in each window type is the Printer Setup option. When selected, this option allows you to select an alternative printer destination if one exists; it also allows limited formatting of the information sent to the printer.

Try It

Print the graph you produced in the previous Try It! exercise.

Application 1: **OPTICAL FIBER PROPAGATION DELAY**

Electrical Engineering

Users of computer networks want the greatest speed possible. The speed with which a network operates is a function of the speed of the network nodes, the links that interconnect the nodes, and the software protocols that manage the movement of data through the links. In this application problem, we will focus on a link that connects two LANs, which together form a simple WAN.

The primary characteristics of a link are its bandwidth, impedance, and propagation delay. Bandwidth, in Mbps (millions of bits per second), is a measure of a link's ability to transport data; in effect, it is the size of the "pipe" that carries the data. Impedance attenuates the signal; therefore, it sets the maximum length of the link. The greater the attenuation, the shorter the link must be in order to avoid loss of the signal. In a typical WAN, barring congestion problems, the propagation delay makes up the majority of the response time. The response time is the time it takes to receive a response from the node at the other end of the link once a

message has been sent to it. We will assume the response time is twice the propagation delay of the link, that is, the time it takes a signal to propagate from one end of the link to the other end and back.

1. Define the Problem

The problem is to minimize the response time between two LANs. The two LANs are separated by 10 km. The link between them must be capable of carrying at least 500 Mbps. The propagation delay through the link must be calculated so that response time is known.

2. Refine the Problem Definition

The most common ways to implement computer network links are to use twisted pair cable, coaxial cable, optical fiber cable, telephone lines, microwave radio signals, or satellite radio signals. We must collect data on each of these possibilities. The data will include the bandwidth, attenuation, and propagation delay characteristics of each. These characteristics are provided, as appropriate, by cable manufacturers, telephone companies, and government agencies.

3. Research Potential Solutions

We cannot use twisted pair or coax in this implementation because their attenuation characteristics restrict their length to about 1 km, and we must span 10 km. Telephone lines do not have the required bandwidth. Microwave and satellite technology are too expensive.

We decide to use optical fiber cable. It has low attenuation, allowing lengths of tens of kilometers, and a bandwidth of over 1 Gbps (1,000 Mbps). Its cost is very low compared to microwave and satellite communication. We know the bandwidth and attenuation characteristics well.

4. Choose and Implement a Solution

Now, we must compute the response time. Binary data is carried as pulses of light within optical fiber cables. Depending on the way in which the light pulses reflect off the inside of the fiber as it propagates, the effective propagation speed of the pulses will be somewhat less than the speed of light. We will assume the propagation speed of signals in our link is 80 percent of the speed of light. Given that the speed of light is 3×10^8 m/s, we can use MATLAB to calculate the response time.

```
» c=3e8;
» speed=0.8*c;
» delay=10000/speed;
» response_time=2*delay

response_time=
  8.3333e-005
```

5. Test the Solution

You have found that the response time of a 10-km optical fiber link is 83.3 microseconds. Testing this solution is actually quite simple since network test equipment is readily available that can measure response times. Therefore, once the link has been implemented, the actual response time can be measured. The actual response time would include the time it takes to transmit and receive a message and the time it takes the receiving node to decide what its response should be.

What If

How might you implement the link in this application problem if the two networks were 1 km apart? If they were 5,000 km apart?

SUMMARY

This chapter covered the basics of using MATLAB, giving you a brief introduction to those MATLAB concepts that will be emphasized in the remainder of this module. Section 2-1 explained how to run and quit MATLAB, described the MATLAB command window, and described the Help feature. Section 2-2 demonstrated how to create and use scalars, variables, expressions, matrices, and vectors within MATLAB. It also showed how to get information about the workspace and introduced the creation of graphs. Section 2-3 explained how to create MATLAB programs, use input and output commands within them, and read and write MATLAB data files.

Key Words

breakpoint
built-in
command window
concatenation
editor/debugger window
expression
figure window
matrix
M-files
menu bar
scalar
scientific notation
shortcut key operation
title bar
tool bar
vector
workspace

MATLAB Special Characters

The following special MATLAB characters were introduced in this chapter:

»	Prompts the user for the next command
;	Suppresses MATLAB responses and separates commands or rows in a matrix
[ ]	Form a matrix or delete a row or column
( )	Contain the indices of a matrix element or the arguments of a function
,	Separates elements of a matrix row or matrix subscripts
:	Delimits starting and ending values in a range of values and creates a vector from a matrix

%	Precedes a comment in a MATLAB program
\n	Used in a *fprintf* function to generate a carriage return in MATLAB output

MATLAB Commands

clc	Clears the command screen
clear	Clears the workspace memory
CONTROL *c*	Causes an abort of a MATLAB command
dbtype	Lists an M-file with line numbers
demo	Runs the MATLAB demonstration program
echo off	Disables echoing of M-file commands
echo on	Enables echoing of M-file commands
edit	Edits an M-file
exit	Exits MATLAB
format	Formats numerical output
help	Accesses help in general or help on a specific M-file
helpdesk	Calls the Help Desk
helpwin	Calls the Help Window
hold on	Freezes plots in the figure window
hold off	Unfreezes plots in the figure window
home	Moves cursor to home position, similar to *clc*
intro	Calls an introductory MATLAB tutorial program
keyboard	Sets a breakpoint in a program, allowing user control
load	Loads a saved workspace
pause	Halts a MATLAB program until the user hits ENTER
quit	Quits MATLAB
return	Returns from user control to program control (see *keyboard*)
save	Saves the workspace, storing all declared variables and their values
type <filename>	Lists an M-file
what	Lists all MATLAB-related files stored in the default directory
whatsnew matlab	Lists the new features of MATLAB 5
which <filename>	Locates <filename> on your default drive
who	Lists the variables stored in the workspace memory
whos	Lists the workspace variables and their sizes

MATLAB Variables and Constants

ans	Default MATLAB variable that stores calculation results
pi	Default MATLAB variable that stores the value of π

MATLAB Functions

cos(x)	Returns the cosine of x
disp(x)	Displays the value of x, which can be text or numbers
exp(x)	Returns the value of e^x
fprintf(format, variables)	Prints formatted information, a C command

grid	Places a grid pattern on a plot
input('text')	Returns user input from the keyboard; the text is displayed
linspace(x,y,z)	Creates a vector that contains *z* equally spaced elements from *x* to *y*
loglog(x,y)	Creates a log-log plot
plot(x,y)	Plots a linear X-Y graph of the data in **x** versus the data in **y**
semilogx(x,y)	Creates a semilog plot with a logarithmic X-axis
semilogy(x,y)	Creates a semilog plot with a logarithmic Y-axis
sin(x)	Returns the sine of *x*
size(x)	Prints the size of *x* in bytes
sqrt(x)	Returns the square root of *x*
title('title')	Adds '*title*' to the current X-Y graph
xlabel('label')	Adds '*label*' to the X-axis of the current X-Y graph
ylabel('label')	Adds '*label*' to the Y-axis of the current X-Y graph

Problems

1. Determine the value of $(1.7 - 3.4)/(5.2 \cdot 8.2)$.
2. Determine the square root of 5.7.
3. Determine the surface area and volume of a cube with a side length of 4.5.
4. Determine the circumference and area of a circle with a radius of 20.
5. Determine the radius of a sphere with a volume of 23.4.

In each of Problems 6 through 9, enter two equations: the first to define the value of *x*, the second to determine the value of *y*.

6. Determine the value of *y* when $y(x) = 2x - 3.3$ and $x = 3.4$.
7. Determine the value of *y* when $y(x) = e^x + x$ and $x = 4$.
8. Determine the value of *y* when $y(x) = \cos(x + 2)$ and $x = 3.4$.
9. Determine the value of *y* when $y(x) = 10e^{-5x}\sin(\pi x)$ and $x = 3.2$.
10. Create a three-element matrix that includes the names John, Mary, and Ellen. Create a second three-element matrix that contains the addresses 123 F St, 14 Elm Ave, and 1234 C Blvd. Combine the two matrices into one matrix and display the result, listing the names in column one and the addresses in column two. Determine the size of the resulting matrix.
11. Create a three-row table in which the first row contains the range 0 to 2π in increments of $\pi/4$. Each value in the second row is the cosine of the corresponding value in the first row. Each value in the third row is the absolute value of the corresponding value in the second row. Label the rows appropriately. (Note that text row labels cannot be included in the data matrix.)

12. Given the quadratic equation

$$Y = AX^2 + BX + C$$

write a program that has the user provide the values for the coefficients A, B, and C and then uses the following equation to determine the roots of the equation:

$$\text{ROOTS} = \frac{-B \pm \sqrt{B^2 - 4AC}}{2A}$$

13. Create an X-Y graph of the function $y(x) = 5x - 3$ for x equals -10 to +10 in increments of 0.1.
14. Create an X-Y graph of the function $y(t) = 14e^{-5t}$ for t equals 0 to 2 in increments of 0.01.
15. Plot the contents of a vector that is defined by

$$\mathbf{X}_i = \sin(2\pi i/50),\ i = 0..100$$

against the contents of a vector that is defined by

$$\mathbf{Y}_i = \cos(6\pi i/50),\ i = 0..100$$

Use the five-step problem-solving process while working problems 16 through 20.

16. Assume a free-falling object that begins its fall at 1,000 feet above the ground. Ignore air resistance. Create two vectors. The elements of the first vector, representing time in seconds, are the values from 0 to 20 in increments of 0.1. The elements of the second vector contain the vertical position of the falling object at the corresponding times listed in the first vector. Plot the second vector versus the first.
17. Write a program that determines the propagation delay of a signal traveling the length of a coaxial cable. The length of the cable is provided by the user. Assume that the propagation speed of a signal in the cable is 84 percent of the speed of light.
18. Write a program that determines how long it will take an object with no air resistance to hit the ground falling from a height, given in meters, that is provided by the user.
19. Included with MATLAB are numerous MAT-files. Some of these files contain data that represent digitized sounds. One such MAT-file is SPLAT.MAT, which contains the matrices **Fs** and **y**. Read SPLAT.MAT and plot **y** versus the indices of the data it contains. To create an array of indices, create a vector **x** with elements that range from 1 to the number of elements in **y**. Then use *plot(x,y)*. Do the same with the files GONG.MAT and CHIRP.MAT. You can print these graphs by selecting Print in the File menu of the figure window.
20. Write a program that reads the MAT-file SPLAT.MAT, asks the user how many data points of the digitized sound to plot (up to 10,001, the size of matrix **y**, which contains the digitized data), and then plots the data.

3 Matrices and Programming

Microprocessors

The central processing unit (CPU) of a digital computer consists of the arithmetic-logic unit, register unit, and control unit. Every CPU is designed to execute a finite set of instructions, called the instruction set. Each instruction of that set, when executed under the direction of the control unit, performs some small program task, such as reading data from memory or adding the contents of two registers.

The CPU designer must decide how many instructions to include in the instruction set, as well as the power of each instruction. Including too few or too simple instructions makes it difficult to write system programs, such as compilers. Including too many or too powerful instructions complicates the control unit, increasing its cost and its size. In the application problem at the end of this chapter, we will show how MATLAB can be used to help create the specifications of an instruction set for a CPU.

INTRODUCTION

This chapter begins by giving an overview of MATLAB's large set of built-in functions. A discussion of array operations, which involve element-by-element arithmetic, follows. Then MATLAB's built-in special matrix functions are described. An explanation of how to write your own functions completes the first section of the chapter.

Section 3-2 shows you how to extend the power of MATLAB programs through the use of iterative loops. It describes the *if, for, while,* and *break* commands. Section 3-3 covers the matrix operations: transpose, multiplication, inverse, power, and determinant. The emphasis in this section is on creating your own programs and functions.

3-1 FUNDAMENTALS

Built-In Functions

MATLAB derives much of its power from its extensive set of functions. Most of these functions are built into the MATLAB program; the others are implemented in M-files provided with MATLAB. To these, you can add your own M-file functions. Also available are powerful, specialized M-file functions that can be purchased from MathWorks, as well as user-written M-file functions that are public domain software. Conveniently, all functions that are defined by M-files are used in exactly the same way as are the built-in functions.

The general categories of MATLAB functions include

- Elementary math functions
- Specialized math functions
- Matrix functions
- Data analysis functions
- Graphing functions
- Polynomial functions
- Numerical methods functions
- Graphical user interface functions
- File I/O functions

We will address, at least partly, all but the last of these topics in the remainder of this module. This chapter covers elementary functions, array operations, special functions, and matrix operations.

Elementary math functions include the arithmetic functions, trigonometric and hyperbolic functions, absolute value, square root, log, round, and many other similar functions. The elementary functions are *built-in functions,* implemented within the executable code of MATLAB.

Specialized math functions perform advanced operations such as *bessel, beta, gamma,* and computing a cross product. They also create certain special matrices, such as the Pascal and magic matrices. Special functions are usually implemented in user-readable M-files.

Matrix functions perform operations such as transposition, exponentiation, and the determination of determinants. Matrix operations tend to be more complex than those performed by array operations. Chapter 4 covers matrix functions, numerical methods, data analysis, and graphing.

Data analysis functions include functions that find a mean and standard deviation. *Graphing functions* create 2-D, 3-D, and 4-D plots. *Polynomial functions* include functions that find roots and fit a polynomial to a set of data. *Numerical methods functions* include functions that determine derivatives and integrals of functions. *Graphical user interface (GUI) functions* create custom windows-based user interfaces. The *file I/O functions* include the file-handling functions, such as *fopen*, *fclose*, *fread*, and *fwrite*. There are other MATLAB function categories; to get a listing of these categories and learn more about the functions in them, enter the command *helpwin*.

Operations performed on matrices fall into one of two categories: array operations or matrix operations.

Array operations designate functions, elementary or more advanced, that perform element-by-element mathematical operations on vectors and matrices. For example, cubing each individual element in a matrix is an array operation. An array operation is distinguished by a period preceding the operator. For example, you would cube each element in the matrix **A** by typing

```
» A=A.^3
```

Matrix operations work on matrices as whole entities. Examples include finding the inverse, determinant, or dot product of a matrix and multiplying or dividing two matrices. For example, you would cube the entire matrix **A**, by typing

```
» A=A^3
```

The matrix **A** would have to be a *square matrix*, that is, a matrix with the same number of columns as it has rows.

Elementary Functions

All of MATLAB's elementary functions, including addition, subtraction, and all the common trigonometric and hyperbolic functions, operate on scalars, vectors, and matrices in the same way. Scalars are easy enough. The following example computes a sum, a cosine, a hyperbolic sine, and an absolute value:

```
» 2+3

ans=

  5

» cos(pi/4)

ans=

  0.7071

» sinh(pi)

ans=

  11.5487

» abs(-2)

ans=

  2
```

Elementary functions operate on vectors element by element. For example, consider the following five-element vectors:

```
» A=[1 2 3 4 5];
» B=[2 3 4 5 6];
» A+2

ans=

  3   4   5   6   7

» A+B

ans=

  3   5   7   9   11

» C=[pi/8 pi/4 pi/2 pi];
» D=cos(C)

D=

  0.9239   0.7071   0.0000   -1.0000
```

Elementary functions also operate on matrices element by element. For example, to find the square root of the absolute value of each element in the matrix B,

```
» B=[-2 -1 -1/2 0;  2  1  1/2  0];
» B=sqrt(abs(B))

B=

  1.4142   1.0000   0.7071   0
  1.4142   1.0000   0.7071   0
```

Other commonly used elementary MATLAB functions include

exp(x) Returns the value of e^x

round(x) Returns the value of the integer nearest **x**

fix(x) Returns the value of the integer between **x** and 0 that is nearest **x**

floor(x) Returns the value of the integer nearest to and less than **x**

ceil(x) Returns the value of the integer nearest to and greater than **x**

log(x) Returns the value of the natural logarithm of **x**

log10(x) Returns the value of the log base 10 of **x**

sign(x) Returns -1 if **x** is less than 0, 0 if **x** is equal to 0, and +1 if **x** is greater than 0

rem(x,y) Returns the value of the remainder of the integer division **x./y** (an array operation, as indicated by the period preceding the divide symbol)

EXAMPLE 3-1

Using Elementary Functions

Create a vector **y** that contains 11 evenly spaced values, rounded to the nearest tenth, of the function *cos(x)* for *x* equals 0 to π. Remember that the *linspace(x,y,n)* function creates a row vector that has *n* evenly spaced elements that range from *x* to *y*.

SOLUTION

As you solve the example problems of this chapter, remember to press (ENTER) at the end of each command line and to use the *clear* command as needed to clear out all past variables before starting each new problem. Solving this problem one step at a time, we have

```
» y=linspace(0,pi,11);
» y=cos(y);
» y=round(10*y)/10

y =
  Columns 1 through 7
   1.0000  1.0000  0.8000  0.6000  0.3000  0   -0.3000
  Columns 8 through 11
  -0.6000  -0.8000 -1.0000 -1.0000
```

When there are more columns in an output than can fit on a single line, MATLAB prints as many columns as will fit on each line, labeling each line of output appropriately. In this example, the two lines of output were labeled "Columns 1 through 7" and "Columns 8 through 11."

Try It Determine the values of e^1, e^2,..., e^{11}, rounded to the nearest integer.

Array Operations

As we noted, the term *array operations* refers to element-by-element arithmetic operations (multiplication, division, and exponentiation) on vectors and matrices. When an array operation is desired, a period must precede the arithmetic operator. For example, in the following illustration, you would multiply each element in the five-element vector **x** by the corresponding element in the five-element vector **y**, placing the results in vector **z**:

```
» x=[1 2 3 4 5];
» y=[2 3 4 5 6];
» z=x.*y

z =
  2  6  12  20  30
```

Obviously, two or more arrays or matrices must be the same size in order to perform array (that is, element-by-element) operations on them.

As another example, let us determine the square of a matrix using both the array operator and the matrix operator. First, create the matrix.

```
» x=[1 2 3; 1 2 3; 1 2 3]

x=

     1     2     3
     1     2     3
     1     2     3
```

Then create the square of matrix **x** using an array operation. The carat character (^) is used in MATLAB to designate exponentiation. Note that each individual element is squared.

```
» y=x.^2

y =

     1     4     9
     1     4     9
     1     4     9
```

Lastly, create the square of the matrix **x** using a matrix operation. Matrix multiplication is explained later in this chapter.

```
» z=x^2

z =

     6    12    18
     6    12    18
     6    12    18
```

EXAMPLE 3-2

Using Array Operations

Create the 4 × 4 matrix $\mathbf{y} = \mathbf{x}^2/2$ on an element-by-element basis when

$$\mathbf{x} = \begin{matrix} 1 & 2 & 3 & 4 \\ 2 & 3 & 4 & 5 \\ 3 & 4 & 5 & 6 \\ 4 & 5 & 6 & 7 \end{matrix}$$

SOLUTION

First, create the 4×4 matrix **x**.

```
» x=[1  2  3  4; 2  3  4  5; 3  4  5  6; 4  5  6  7];
```

Using the carat character (^) to designate exponentiation and the period to designate an array operation, type

```
» y=x.^2./2

y=

   0.5000    2.0000     4.5000     8.0000
   2.0000    4.5000     8.0000    12.5000
   4.5000    8.0000    12.5000    18.0000
   8.0000   12.5000    18.0000    24.5000
```

Try It Create the vector $V = e^{-2t} \cos(\pi t/8)$ on an element-by-element basis when t varies from 0 to 1 in 0.05 increments. Check your answers with a calculator.

Special Matrices

MATLAB includes a number of functions that either perform special operations on matrices or generate *special matrices.* As mentioned earlier, some of those functions that perform special operations are *bessel, beta,* and *gamma.* The more useful of those functions that generate special matrices are

zeros(m,n)	Returns an $m \times n$ matrix of all zeros
ones(m,n)	Returns an $m \times n$ matrix of all ones
eye(n)	Returns an $n \times n$ identity matrix
magic(n)	Returns an $n \times n$ magic matrix
pascal(n)	Returns an $n \times n$ Pascal's triangle
hadamard(n)	Returns an $n \times n$ Hadamard matrix

An *identity matrix* is a square matrix of all zeros except that the diagonal from the upper left to the lower right is all ones. Its name, *identity matrix,* comes from the fact that, when a matrix **X** is multiplied by an identity matrix, the product is **X**.

A *magic matrix* is a square matrix in which the rows, columns, and diagonals all add to the same number. For example, the rows, columns, and diagonals each add to 15 in a 3×3 magic matrix.

```
» magic(3)

ans=

   8  1  6
   3  5  7
   4  9  2
```

The *pascal(n)* function returns an $n \times n$ matrix that contains a *Pascal's triangle*. The Pascal's triangle is formed by the lower left to upper right diagonal and those elements above and to the left of that diagonal. The diagonal contains the coefficients of a binomial expansion of the form $(a + b)^{n-1}$. For example,

```
» pascal(4)

ans=

   1   1   1   1
   1   2   3   4
   1   3   6  10
   1   4  10  20
```

generates the coefficients for the polynomial expansion of $(a + b)^{n-1}$, where $n = 4$.

The coefficients are in the lower left to upper right diagonal (1 3 3 1); therefore, $(a + b)^3 = 1a^3 + 3a^2 + 3ab^2 + 1b^3$.

The *hadamard* function returns an $n \times n$ Hadamard matrix. Hadamard matrices have application to areas that include signal processing and numerical analysis. They are matrices of -1s and 1s whose columns are orthogonal; therefore, the following is true:

$$\mathbf{H}'*\mathbf{H} = n*\mathbf{I}$$

where **H** is an $n \times n$ Hadamard matrix, **H'** is the transpose of **H**, and **I** is an $n \times n$ identity matrix. The MATLAB command

```
» contour(hadamard(32))
```

produces an interesting plot that is similar to a Navajo blanket design.

EXAMPLE 3-3 Using Special Matrices

Determine the coefficients for the binomial $(a + b)^4$.

SOLUTION

We need a fourth degree expansion, therefore, $n = 5$.

```
» pascal(5)

ans=

   1   1   1   1   1
   1   2   3   4   5
   1   3   6  10  15
   1   4  10  20  35
   1   5  15  35  70
```

Reading the lower left to upper right diagonal, we have the answer: $(a + b)^4 = 1a^4 + 4a^3b + 6a^2b^2 + 4ab^3 + 1b^4$.

Try It

What would be the sum of each of the rows, columns, and diagonals of a 4 × 4 magic matrix?

User-Defined Functions (M-Files)

In Chapter 2, you learned how to write simple MATLAB programs and how to read and write data files. To make your programs more useful, you can write your own functions and include them as subroutines in your programs or use them at the command line in the command window.

As when writing MATLAB (M-file) programs, you use the MATLAB Editor/Debugger to create custom functions. These functions, like MATLAB programs, must be saved in files with a .M extension; therefore, they are also called M-files. User-defined functions can accept zero or more arguments and return zero or more arguments. Following is an example of a user-defined function that has been created using the MATLAB Editor/Debugger and saved in a file called AREA.M:

```
function a=area(r)
a=pi.*r.^2
```

You can use this function to determine the area of a circle with radius *r*.

```
» A=[ 0  1  2; 4.4  5.5  6.6];
» area(A(1,1))

ans=
  0

» area(A(1,:))

ans=
  0   3.1416   12.5664

» area(A)

ans=
   0        3.1416    12.5664
  60.8212   95.0332   136.8478
```

There are certain rules to which you must adhere when you write your own functions.

1. The first line must begin with the word *function*, followed by the output arguments (if any), an equals sign, the name of the function, and the input arguments (if any) within parentheses. For example,

   ```
   function a=area(r)
   ```

2. When there are multiple input arguments, they must be separated by commas. For example,

   ```
   function c=hypot(a,b)
   ```

3. When there are multiple output arguments, they must be listed within a vector, separated by commas. For example,

```
function [distance, velocity, accel]=position(X)
```

   Since the function is returning multiple arguments, it must be used as follows:

```
» [ d  v  a]=position(A)
```

4. All variables used in each function are local. That is, their use has no effect on variables of the same name in the program that references the function.
5. The functions *nargin* and *nargout* can be used within functions to determine the number of input arguments and the number of output arguments, respectively, used in the reference to the function. This allows the programmer to handle the cases when the function uses a smaller number of arguments than that for which it was designed. For example, if you want to return a default value of 5 for the variable *a* if the function reference includes no input arguments, the function would include the following:

```
if nargin==0 a=5;
```

6. Comments should be added to each function, following the first line. The comments should include information on the use of the function and how it works. Program comments aid in debugging and in future modification of the function. Each comment line must start with a percent symbol (%). If the name of a function (or program) is entered following the *help* command, the comment lines will be listed. For example,

```
» help position
```

   will list all the comments in the file POSITION.M.
7. A user-defined function is referenced by the name of the M-file in which it is defined, *not* by the name given the function in the first line of the file. Consider the following function, named *area*, that is saved in the file MYAREA.M:

```
function a=area(r)
a=pi.*r.^2
```

   The following would determine the area of a circle with a radius of 3:

```
answer=myarea(3)
```

   However, MATLAB would respond to the following with an error message:

```
answer=area(3)
```

   To avoid confusion, you should always use the same name for the function and the M-file.

EXAMPLE 3-4

Writing a User-Defined Function

Write a function that takes the mass at rest and the linear speed in meters per second of a particle as its input arguments, and returns the present mass and momentum of the particle.

SOLUTION

First, open the editor/debugger window by selecting M-file from the New option under the File menu. Type the following in the editor/debugger window:

```
function [mass, momentum]=massmom (m0, speed)
% This function computes relative mass & momentum, given a
velocity in m/s.
% It uses array operations (element-by-element operations)
mass=m0./sqrt(1-speed.^2 ./3e8^2);
momentum=mass.*speed;
```

Note that only the last two lines in the function are MATLAB commands and that both end with a semicolon. Omitting the semicolons would result in MATLAB echoing the results of each command whenever the function is used. Always be sure to include these semicolons in your functions unless you want the results echoed; otherwise, the user may be confused, wondering from where those often intermediate results came.

Now that you have finished writing your function, select Save As under the File menu in the editor/debugger window, and save the file as MASSMOM.M. Select Exit Editor/Debugger under the File menu to close the editor/debugger window, and return to the command window.

The following tests the function by determining the mass (in grams) and momentum (in gram-meters per second) of a 1-gram particle traveling 2.9×10^8 meters per second:

```
» [ma mo]=massmom(1, 2.9e8)

ma=

  3.90566732942472

mo=

  1.132643525533167e+009
```

If any errors occurred, select Open under the File menu, then double-click on MASSMOM.M to edit the function.

Try It

Write a function that takes as its input arguments the lengths of the two sides of a right triangle and returns the length of the hypotenuse.

3-2 PROGRAM CONTROL FLOW

if Command

As with programs written in any other language, MATLAB programs often require the use of an *if* command. As an example, the following user-defined function performs the opposite operation as that performed by *abs*, which returns the absolute value of its argument. Note that, as you press ENTER at the end of each line, the MATLAB Editor/Debugger automatically inserts tabs as appropriate.

```
function n=nabs(a)
if a>0
  n=-a;
else
  n=a;
end
```

The available *relational operators* are as follows:

<	Less than
>	Greater than
<=	Less than or equal to
>=	Greater than or equal to
==	Equal to
~=	Not equal to

Also available are the following *logical operators*:

&	AND
|	OR
~	NOT

You can nest *if* commands within *if* commands. MATLAB also provides an *elseif* command, which we will use in Example 3-5.

EXAMPLE 3-5 Using the *if* Command

Write a function that takes a single input argument and then reports, in text, whether that argument is scalar, vector, or matrix. No argument is returned.

SOLUTION

Open the editor/debugger window and type

```
function test(x)
% This function determines whether x
% is a scalar, vector or matrix
[m n]=size(x);
if m==n & m==1
  disp('It is a scalar')
elseif m==1 | n==1
  disp('It is a vector')
else
  disp('It is a matrix')
end
```

Note that this function defines no output arguments. Save the function as TEST.M, close the editor/debugger window, and test the function in the command window.

```
» a=2;b=[2 3];c=[1 2; 3 4];
» test(a)

It is a scalar

» test(b)

It is a vector

» test(c)

It is a matrix
```

Try It

Write a function that accepts only a single scalar input argument and reports, in text, whether the argument is less than zero, equal to zero, or greater than zero. Report an error message if the argument is not a scalar. No argument is returned.

---

for Loops

The *for* command is used to execute a command, or set of commands, a predetermined number of times. The form of a *for* loop is

```
for i=range
  commands;
end
```

As when creating *if-then-else* constructs, the MATLAB Editor/Debugger automatically inserts tabs as appropriate in *for*, and other, loops.

The number of times the loop is executed is the same as the number of values the range variable (*i* in this case) assumes. The following *for* loop would be executed ten times:

```
for j=1:10
  M(j)=j;
end
```

The default increment value of the range variable is 1; however, it can be negative and need not be an integer. For example, the following begins a *for* loop that would be executed 101 times:

```
for k=10:-0.1:0
```

When a *for* loop completes execution, the value of its range variable is the terminal value of its range. As with *if* commands, *for* commands can be nested. As an example, the following fills a 10×10 matrix **X** with 2s:

```
for i=1:10
  for j=1:10
    X(i,j)=2;
  end
end
```

EXAMPLE 3-6 Using *for* Loops

Write a function that searches a matrix input argument for the element with the largest value and returns the indices of that element. Remember to not insert tabs; the Editor/Debugger will do this for you.

SOLUTION

Open the editor/debugger window and type

```
function [a, b]=largest(x)
% This function finds the indices
% of the largest element in x
[m n]=size(x);
largest=x(1,1);
a=1; b=1;
for i=1:m
  for j=1:n
    if x(i,j)>largest
      a=i;
      b=j;
```

```
    end
  end
end
```

Save the function as LARGEST.M and close the editor/debugger window. Test the function.

```
» a=2;b=[2 3];c=[1 2; 3 4];
» [row column]=largest(a)

row =
  1
column =
  1

» [row column]=largest(b)

row =
  1
column =
  2

» [row column]=largest(c)

row =
  2
column =
  2
```

Remember, the results indicate the row and column indices of the largest element in the parameter passed to the function.

Try It

Write a function that searches a matrix input argument for the element whose value is closest to zero. The function returns the value and its indices.

---

while Loops and *break*

Where the *for* command allows a group of commands to be executed a predetermined number of times, the *while* command allows the execution of a set of commands an indefinite number of times. The form of a *while* loop is

```
while range
  commands;
end
```

A *while* loop is often used to repeat a process as many times as the user desires; the loop terminates when the user enters a terminating value. The *break* command is often useful in a *while* loop. Example 3-7 demonstrates the principle.

EXAMPLE 3-7

Using *while* Loops

Write a program that asks the user to enter the scalar values for a, b, c, and x and then returns the value of $ax^2 + bx + c$. The program repeats this process until the user enters zero values for all four variables.

SOLUTION

Open the editor/debugger window and type

```
function quadra
% This program returns the value of the
% quadratic equation: ax^2+bx+c
disp('I will evaluate ax^2+bx+c.')
disp('If you enter a, b, c, and x.')
a=1; b=1; c=1; x=1;
while a~=0 | b~=0 | c~=0 | x~=0
  disp('Enter 0, 0, 0, and 0 to quit.')
  a=input('Please enter the value of a: ');
  b=input('Please enter the value of b: ');
  c=input('Please enter the value of c: ');
  x=input('Please enter the value of x: ');
  if a==0 & b==0 & c==0 & x==0
    break
  end
  answer=a*x^2+b*x+c;
  fprintf('%f is the answer.\n',answer)
end
```

Save this program as QUADRA.M, and close the editor/debugger window. Run and test the program.

```
» quadra
```

Try It

Modify QUADRA.M so that it checks each input and breaks if any input is not a scalar. Note that the size of a scalar is 1×1.

3-3 MATRIX OPERATIONS

Matrix operations differ from array operations when performing multiplication, division, and exponentiation. If you want an array (element-by-element) operation performed, you must precede the arithmetic operator with a period. When performing matrix operations, no period is used. We will describe the rules for performing matrix multiplication, division, and exponentiation after the following discussion of the matrix transpose.

Transpose

The transpose of a matrix swaps its rows and columns. Row 1 becomes column 1, column 1 becomes row 1, and so on for the other rows and columns. The transpose operation therefore transforms a row vector into a column vector and vice versa.

The apostrophe (') is used to designate the transpose operation as follows:

```
» A=[1  3  7  11; 2  5  9  13; 3  6  10  16]

A=
   1   3   7  11
   2   5   9  13
   3   6  10  16

» B=A'

B=
    1   2   3
    3   5   6
    7   9  10
   11  13  16
```

You may find it convenient to create column vectors by defining them as the transpose of a row vector. For example,

```
» A=[1  2  3]'

A=
   1
   2
   3
```

Column vectors are often combined to form tables.

```
» a=[1  2  3]';
» b=cos(a);
» c=[a  b]
```

```
C=
   1.0000    0.5403
   2.0000   -0.4161
   3.0000   -0.9900
```

EXAMPLE 3-8

Using the Transpose Function

Determine whether an element-by-element product of a 3 × 3 magic matrix and its transpose is a magic matrix.

SOLUTION

First, create a matrix **D** that is the product of a 3 × 3 magic matrix and its transpose.

```
» D=magic(3).*magic(3)'

D=
   64    3   24
    3   25   63
   24   63    4
```

Now, compute the sums of columns.

```
» sum(D)
```

The *sum* function adds the elements in each column in its matrix input argument, returning a row vector.

```
ans=
   91   91   91
```

You can see by inspecting **D** that its rows are identical to its columns; therefore, the rows, too, each add to 91. So far, so good. Now, add the upper left to lower right diagonal elements.

```
» D(1,1)+D(2,2)+D(3,3)

ans=
   93
```

We have just proved that the product of a 3 × 3 magic matrix and its transpose is *not* a magic matrix.

Try It

Determine whether the transpose of a 3 × 3 magic matrix is a magic matrix. Determine whether the transpose of an identity matrix is an identity matrix.

Matrix Multiplication

The symbol * denotes multiplication of whole matrices (versus element-by-element multiplication). The matrix multiplication **X*Y** can be performed only when the second (column) dimension of **X** is the same as the first (row) dimension of **Y**. That is, an $m \times n$ matrix can be multiplied only by an $n \times p$ matrix. The result will be an $m \times p$ matrix.

A row vector and a column vector can be multiplied only if they have the same number of elements. In the following example, a 1×4 vector is multiplied by a 4×1 vector; the result is a 1×1 matrix (that is, a scalar):

```
» x=[1  2  3  4]

x =
   1  2  3  4

» y=[1; 2; 3; 4]

y =
   1
   2
   3
   4

» P= x*y

P=
   30
```

In general, each element P(i,j) in **P=x*y** is computed as the sum of the products of row i in **x** and column j in **y**. Since **x** is a 1×4 matrix and **y** is a 4×1 matrix, the result is a 1×1 matrix. Its only element P(1,1) is computed as the sum of the products of row 1 of **x** and column 1 of **y**, that is, as A(1,1) * B(1,1) + A(1,2) * B(2,1) + A(1,3) * B(3,1) + A(1,4) * B(4,1). Example 3-9 demonstrates the process.

EXAMPLE 3-9

Using Matrix Multiplication

Multiply 3×2 matrix **x** by 2×4 matrix **y**.

```
x = 1  2      y = 2  3  4  5
    2  3          3  4  5  6
    3  4
```

SOLUTION

First, define the two matrices.

```
» x=[1  2; 2  3; 3  4];
» y=[2  3  4  5; 3  4  5  6];
```

Now, compute the product.

```
» p=x*y

p =

   8   11   14   17
  13   18   23   28
  18   25   32   39
```

Note that p(1,1) equals the sum of the products of row 1 of **x** and column 1 of **y**; that is, p(1,1) = x(1,1) * y(1,1) + x(1,2) * y(2,1) = 1 * 2 + 2 * 3 = 8.

Any matrix can be multiplied by its transpose because the dimensions of the transpose of an $m \times n$ matrix are $n \times m$. Multiplication is defined for both **A*****A**' and **A**'***A**. For example,

```
» A=[0  1; 1  2; 2  3]

A=

  0  1
  1  2
  2  3

» A*A'

ans=

  1  2   3
  2  5   8
  3  8  13

» A'*A

ans=

  5   8
  8  14
```

If **A** is not a square matrix, **A*****A**' and **A**'***A** do not have the same dimensions. If **A** is an $n \times n$ square matrix, **A*****A**' and **A**'***A** will both be $n \times n$ matrices; however, they will not, in general, be the same matrix. Therefore, the operation is not commutative.

Related to the matrix multiply operation is the operation, matrix division operation, which is well applied to solving sets of linear equations. We will discuss matrix division in Chapter 4.

Try It

Is matrix multiplication commutative? That is, in general, does **A*****B**=**B*****A**? Why or why not?

Under what conditions can a matrix be squared, that is, multiplied by itself?

Matrix Inverse

The inverse function is defined only for square matrices. By definition, the inverse of a square matrix is defined as the matrix $\mathbf{A}^{-1}$ such that the products $\mathbf{A}*\mathbf{A}^{-1}$ and $\mathbf{A}^{-1}*\mathbf{A}$ both equal the identity matrix. For example, in the following, we see that **A** and **B** are inverses of each other:

```
» A=[1  2; 3  4]

A=

  1  2
  3  4

» B=[-2 1; 1.5 -0.5]

B=

  -2.0000    1.0000
   1.5000   -0.5000

» A*B

ans=

  1  0
  0  1

» B*A

ans=

  1  0
  0  1
```

There are two ways to create a matrix that is the inverse of another matrix.

```
» B=inv(A)

B=

  -2.0000    1.0000
   1.5000   -0.5000

» B=A^(-1)

B=

  -2.0000    1.0000
   1.5000   -0.5000
```

EXAMPLE 3-10

Using the Inverse Function

Write a program that receives a matrix as an input parameter. The program then checks to make sure the matrix is square; if it is square, the program reports the inverse of the matrix. If the matrix is not square, the program reports an error message.

SOLUTION

Open the editor/debugger window and type

```
function myinv(x)
% This program, myinv, returns the inverse of an
% input matrix ONLY if that matrix is square.
[m n]=size(x);
if m==n
disp('The inverse is:')
inv(x)
else
disp('The matrix is not a square one!')
end
```

Save the program as MYINV.M, and close the editor/debugger window. Test the program on both a square and a nonsquare matrix.

Try It

Rewrite the program in the previous example as a function to which a matrix is passed as an argument.

Matrix Powers

Recall that the array operation **A**.^2 squares each individual element in the matrix **A**. The matrix can have any number of rows and columns. However, the matrix operation **A**^2 is equivalent to the matrix multiplication **A*****A**. Since the matrix multiplication **X*****Y** can be performed only when the second dimension of **X** is the same as the first dimension of **Y**, clearly **A** must be a square matrix in order to square it. It follows, then, that square matrices, and only square matrices, can be raised to any power.

EXAMPLE 3-11

Using Matrix Powers

Write a function that performs the matrix operation $\mathbf{y}^2 + \mathbf{y} + 1$ for any square matrix **y** passed to it.

SOLUTION

Open the editor/debugger window and type

```
function sum=quadra(y)
% This function computes the quadratic y^2+y+1
% for any square matrix y passed to it.
sum=y^2+y+1;
```

Save the function as QUADRA.M, and close the editor/debugger window.

An example of using this function is

```
» C=[ 1 2; 3 4]
» quadra(C)

ans=
    9  13

   19  27
```

Try It Write a function that takes a matrix as its input argument. If the matrix is a square matrix, the function asks the user to enter the power to which he or she wants the matrix raised. The function returns the matrix raised to the requested power.

Determinants

Determinants are useful in a number of areas of engineering, especially when solving simultaneous equations. The determinant of a matrix is a scalar computed from the elements of the matrix. The operation is defined only for square matrices. For a 2×2 matrix **X**, the determinant is

$$|\mathbf{X}| = \left|\begin{bmatrix} X(1,1) & X(1,2) \\ X(2,1) & X(2,2) \end{bmatrix}\right| = X(1,1) \cdot X(2,2) + X(1,2) \cdot X(2,1)$$

For a 3×3 matrix **X**, the determinant is

$$\begin{aligned} |\mathbf{X}| = & [X(1,1)X(2,2) - X(2,1)X(1,2)]X(3,3) \\ & - [X(1,1)X(2,3) - X(2,1)X(1,3)]X(3,2) \\ & + [X(1,2)X(2,3) - X(2,2)X(1,3)]X(3,1) \end{aligned}$$

For example, the determinant of the matrix

$$\begin{matrix} 0 & 1 & 2 \\ 3 & 4 & 5 \\ 6 & 7 & 8 \end{matrix}$$

is

$$[0\cdot 4-3\cdot 1]\cdot 8-[0\cdot 5-3\cdot 2]\cdot 7+[1\cdot 5-4\cdot 2]\cdot 6=(-24)-(-42)+(-18)=0$$

Computing the determinants of larger matrices is a more involved process that should be avoided when working manually. Fortunately, MATLAB includes the matrix operation *det* that does the work for you using a faster, more exotic algorithm than that just described. Example 3-12 demonstrates the use of *det.*

EXAMPLE 3-12

Evaluating a Determinant

Write a function that takes as its input arguments two square matrices and returns the determinant of the first divided by the determinant of the second.

SOLUTION

Open the editor/debugger window and type

```
function answer=detdiv(x,y)
% This function receives two square matrices
% and returns the determinant of the 1st
% divided by the determinant of the 2nd
[m n]=size(x); [p q]=size(y);
if m==n & m==q & m==q
  answer=det(x)/det(y);
else
  disp('Both matrices must be square.')
end
```

Save the function as DETDIV.M, and close the editor/debugger window. Test the function.

```
» a=[1 2 3; 4 5 6; 6 7 9];
» b=[1 3 1; 6 8 6; 3 9 6];
» D=detdiv(a,b)

D=
  0.1000
```

Try It

Write a program that gives the user the following menu of choices of functions:

1 = transpose

2 = inverse

3 = power

4 = determinant

Then ask the user to enter the name of a matrix. If appropriate, make sure the matrix is square. If choice 3 is selected, ask the user for the power to which he or she wants the matrix raised.

Application 2: CPU INSTRUCTION SET DESIGN

Computer Engineering

One of the choices engineers must make in the design of a computer is the specification of the instruction set. In general, the choice of the specific instructions to be included depends on the intended use of the computer. However, tradeoffs are typically necessary. For example, there may not be room on a silicon microprocessor chip to implement all desired instructions. The complexity, and therefore time of execution, of some desired instructions may degrade the CPU performance to an unacceptable level. Ultimately, a CPU design is a compromise between complexity that results in efficient programs and simplicity that results in greater hardware performance.

As with many engineering designs, it cannot be known exactly how a computer will be used until it *is* used. The performance of a computer depends on how it is used, that is, on the types of programs it is used to execute. Therefore, once in use, computers are often monitored to determine those instructions that are used most often and those that are used least often. Later versions of the computer can then be optimized to execute the most used instructions more quickly. Little used instructions may be dropped from the instruction set altogether, thereby simplifying the design.

1. Define the Problem

We want to optimize the design of an existing computer CPU, maximizing its speed. We propose a redesign of the control unit that implements the instruction set.

2. Refine the Problem Definition

We install system software in several examples of the computer that is being used by typical end users. The software monitors the instruction flow from memory to the CPU, counting the instances of execution of each instruction in the instruction set.

The software generates a data file, INST_USE.DAT, which contains one column of data that lists the average number of times each instruction

was executed per minute during the time the CPU was monitored. The first entry is the value for instruction 1, the second entry is the value for instruction 2, and so on for 122 instructions. The following MATLAB session reads the data file and plots a histogram of its contents:

```
» load inst_use.dat
» size(inst_use)

ans=

  122    1

» bar(inst_use)
```

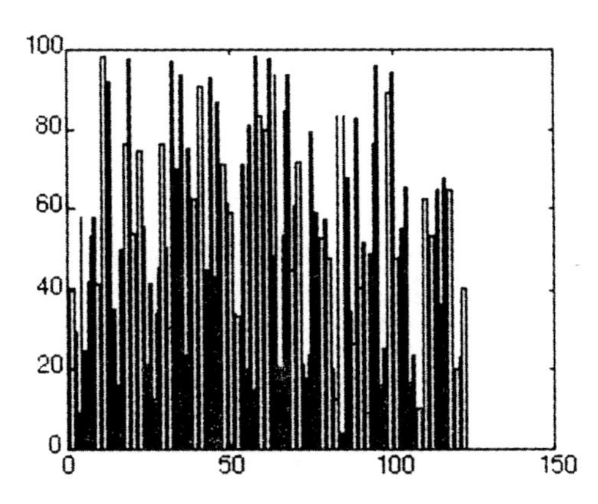

We checked the size of INST_USE just to make sure how much data was in the file. The *bar* command created a histogram, showing how often each of the 122 instructions was executed per minute. (Although histograms are not covered until the next chapter, we introduce the concept here because it is applicable, and it is a simple extension of the X-Y plot introduced in Chapter 2.)

We note from the histogram that some instructions are rarely used, whereas others are often used. Therefore, we decide to redesign the CPU control unit so that it more efficiently executes the instruction set as it is used in actual programs.

3. Research Potential Solutions

From the histogram, we note which instructions are executed most often and which are executed least often. One way to improve the performance of the computer would be to focus on those instructions that are executed most often, redesigning the CPU to execute these instructions more quickly. Another approach would be to eliminate little used instructions from the instruction set. In this way, the CPU design can be simplified and designed to execute the remaining instructions more quickly.

We decide to eliminate those instructions that are used fewer than ten times per minute. The decision is based on experience in improving the performance of CPUs. The following MATLAB *for* loop determines which instructions should be eliminated:

```
» for i=1:122
   if inst_use(i) < 10
     disp(i)
   end
end

     3
    42
    84
    92
    98
   107
   109
   113
   117
   119
```

4. Choose and Implement a Solution

So we decide to eliminate instructions 3, 42, 84, 92, 98, 107, 109, 113, 117, and 119. We redesign the CPU, implementing the remaining 112 instructions. We find that the eliminated instructions were some of the most complex in the instruction set. (Perhaps they were little used because most programmers could not figure out how to use them.) We determine that the clock rate of the simpler CPU can be doubled over that of the older version.

5. Test the Solution

In order to test the solution, we would implement the CPU, building the computer it controls. We would verify that the computer indeed operates dependably at a higher clock rate than that used in the original design. Unfortunately, the software written for the old computer will not, in general, run on the new computer. The old software may include some of the missing ten instructions. However, the source code of old programs can be recompiled to run on the redesigned CPU, which will execute these programs, as well as newly developed programs, faster than did the old CPU.

What If

How would you alter the preceding MATLAB code in order to determine which instructions were executed more than 50 times per minute? Could you write a *for* loop that found the instruction that was executed the most often?

SUMMARY

This chapter focused on three important topics: (1) MATLAB's extensive set of elementary functions, array operators, and matrix operators; (2) writing MATLAB functions; and (3) adding loops to MATLAB programs.

This chapter began with a description of MATLAB's set of elementary functions, such as *exp* and *sqrt*; array operators, which perform element-by-element arithmetic operations; and MATLAB's built-in special matrix functions. An explanation of how to write your own functions followed. Section 3-2 explained how to use iterative loops in your MATLAB programs. The *if*, *for*, *while*, and *break* commands were described. Section 3-3 described the use of a few of the more popular matrix operations: transpose, multiplication, inverse, powers, and determinant. The emphasis in this chapter has been on writing functions and programs.

Key Words

array operation
built-in function
data analysis function
elementary math function
file I/O function
graphics function
GUI function
identity matrix
logical operator
magic matrix
matrix function
matrix operation
numerical method function
Pascal's triangle
polynomial function
relational operator
specialized math function
special matrix
square matrix

MATLAB Special Characters

+   Matrix addition
-   Matrix subtraction
.*   Array (element-by-element) multiplication
./   Array (element-by-element) division
.^   Array (element-by-element) exponentiation
*   Matrix multiplication
/   Matrix division
^   Matrix exponentiation
'   Transpose
<   Less than
>   Greater than
<=   Less than or equal to
>=   Greater than or equal to
==   Equal to
~=   Not equal to
&   Logical AND
| Logical OR
~   Logical NOT

MATLAB Commands

break	Breaks the execution of a program or function
elseif	Offers an alternative in an *if* command
for	Allows execution of a loop a predetermined number of times
if	Allows conditional program branching
while	Allows execution of a loop an indefinite number of times

MATLAB Functions

bar(*x*)	Creates a bar graph of the elements of **x**
ceil(*x*)	Returns the value of the integer nearest to and greater than *x*
det(*x*)	Returns the value of the determinant of the matrix **x**
exp(*x*)	Returns value of e^x
eye(*n*)	Returns an $n \times n$ identity matrix
fix(*x*)	Returns the value of the integer between *x* and 0 that is nearest *x*
floor(*x*)	Returns the value of the integer nearest to and less than *x*
hadamard(*n*)	Returns an $n \times n$ Hadamard matrix
inv(*x*)	Returns the value of the inverse of the matrix **x**
linspace(*x*,*y*,*n*)	Returns a vector that contains *n* evenly spaced elements from *x* to *y*
log(*x*)	Returns the value of the natural logarithm of *x*
log10(*x*)	Returns the value of the log base 10 of *x*
magic(*n*)	Returns an $n \times n$ magic matrix
nargin	Returns the number of input argument in a reference to a function
nargout	Returns the number of output argument in a reference to a function
ones(*m*,*n*)	Returns an $m \times n$ matrix of ones
pascal(*n*)	Returns an $n \times n$ Pascal's triangle
rem(*x*,*y*)	Returns the value of the remainder of the integer division *x./y*
round(*x*)	Returns the value of the integer nearest **x**
sign(*x*)	Returns −1 if *x* is less than 0, 0 if *x* is equal to 0, and + 1 if *x* is greater than 0
sum(*x*)	Returns the sums of the columns of **x**
zeros(*m*,*n*)	Returns an $m \times n$ matrix of zeros

Problems

1. Determine how MATLAB handles
   a. division of a scalar by zero.
   b. division as an array operation (./) when the divisor is a vector or matrix that contains zeros.
   c. division as a matrix operation (/) when the divisor contains some, but not all, zeros.
   d. division as a matrix operation when the divisor contains all zeros.

2. Create a vector **Y** that contains nine evenly spaced values, rounded to the nearest one hundredth, of the function cos(x) for x equals 0 to 2π.
3. Determine the values of $e^1, e^2, \ldots, e^{11}$, rounded to the nearest multiple of 10.
4. Create the matrix $\mathbf{B} = \mathbf{A}^2 + \mathbf{A}/\pi$ on an element-by-element basis when

$$\mathbf{A} = \begin{matrix} 1 & 2 & 3 & 4 \\ 5 & 6 & 7 & 8 \\ 2 & 3 & 4 & 5 \\ 6 & 7 & 8 & 9 \end{matrix}$$

5. Create the vector $\mathbf{V} = e^{-3t/2} \sin(\pi t/8)$ on an element-by-element basis when t varies from 0 to 1 in increments of 0.01.
6. Determine the coefficients of the binomial $(a + b)^5$.
7. Write a function that takes as its input arguments the length of the base and the height of a parallelogram and returns the area of the parallelogram.
8. Write a function that takes as its input arguments the mass at rest and the linear speed in miles per second of a particle, and returns the mass and momentum of the particle at the given speed.
9. Write a function that takes a single input argument and then reports, in text, whether that argument is scalar, vector, square matrix, or non-square matrix. No argument is returned.
10. Write a function that accepts three scalar input arguments and returns their absolute values from smallest to greatest.
11. Write a function that accepts a vector input argument and returns the average of its elements.
12. Write a function that accepts a matrix input argument and returns the average of its elements.
13. Write a function that searches a matrix input argument for the element whose value is farthest from zero. The function returns the value and its indices.
14. Write a function that returns $\mathbf{x}^3 + \mathbf{x}^2 + \mathbf{x}$ (a matrix operation) for any square matrix **x** passed to it.
15. Create a three-column matrix in which the first column contains the range $x = 0{:}\pi/16{:}\pi/2$; the second column contains $\sin(2x)$; and the third column contains $2\sin(x)\cos(x)$. Note that $\sin(2x) = 2\sin(x)\cos(x)$ is a mathematical identity.

   Use the five-step problem-solving process while working Problems 16 through 20.

16. Write a program that asks the user to enter four values, a, b, and c, and x, then returns the value of $ax^2 + bx + c$. The program repeats this process until the user enters the value 0 for each of the four values. Have the program check each input and break if any input is not a scalar.

17. Write a program that asks the user for the name of a matrix the program calls **A** and then asks for the name of a matrix the program calls **B**. The program then reports, in text, whether **A*****B** or **B*****B** or both are defined.
18. Write a program that loads matrix called **data** (previously saved as an ASCII file). The program then checks to make sure the matrix is square. If it is square, the program prints to the monitor screen the inverse and inverse squared of the matrix, with appropriate text to label what is being printed. If the matrix is not square, the program reports an error message. Test your program using the editor/debugger to create and save the named data file.
19. Write a program that asks the user for the beginning, ending, and incremental values for a table in which the three values provided by the user define the values in the first column. The second column lists the log to the base 10 of the corresponding values in the first column.
20. Use the editor/debugger to create an ASCII data file called MYDATA.DAT. Type the following data into MYDATA.DAT:

```
0.1000   1.8905   2.2590   3.1546
0.2000   1.8901   2.2849   3.1403
0.3000   2.0720   2.4415   3.2192
0.4000   2.0822   2.4450   3.1806
0.5000   1.8615   2.4993   3.1694
0.6000   2.0829   2.3498   3.4429
0.7000   1.9734   2.4414   3.1873
0.8000   2.0507   2.4660   3.1277
0.9000   2.0890   2.4874   3.0968
1.0000   1.9904   2.5822   2.9486
```

Columns 2, 3, and 4 represent measurements made by voltage sensors 1, 2, and 3, respectively, at the times specified in the first column. Write a program that reads MYDATA.DAT. The program reports the number of that sensor that has the highest average measurement.

4 Engineering Problems

Electron Tubes

Modern electronics began in the early 20th century with the invention of electron tubes that could store electrical charges and amplify electronic signals. Two types of electron tubes have been developed: vacuum and gas-filled.

Since the invention of the transistor in 1947, semiconductor devices have largely replaced vacuum tubes in many applications. But both types of electron tubes still are widely used in products such as cathode ray tubes (CRTs), X-ray machines, and mercury-arc rectifiers, and most television picture tubes. In the application problem at the end of this chapter, we will look at how MATLAB can be used to analyze a simple electronic circuit.

INTRODUCTION

The three sections of this chapter cover, respectively, matrix functions, data analysis, and plots. Chapter 3 described MATLAB's so-called special functions, which produce unique matrices. In the first section of this chapter, we continue the discussion of matrix functions, focusing on functions that operate on complex numbers, solve nonlinear equations or sets of linear equations, find roots of polynomials, and perform differentiation and integration.

Section 4-2 introduces the subject of data analysis. We begin with an explanation of random number generation. This is followed by a description of common statistical operations, including finding minimums, maximums, means, and standard deviations. The section concludes by showing you how to create histograms.

Section 4-3 briefly describes MATLAB's plotting capabilities. X-Y graphs were introduced in Chapter 2; this chapter briefly describes the creation of simple bar graphs and polar, contour, and 3-D mesh plots.

4-1 MATRIX FUNCTIONS

MATLAB includes several hundred *matrix functions,* and the number is constantly increasing. Most provide complex and specialized operations; however, the most advanced are in the optional toolboxes mentioned in Chapter 1. The built-in matrix functions discussed in this chapter fall into the following general categories:

- Nonlinear (for example, *fmin* and *fzero*)
- Polynomial (for example, *roots* and *polyfit*)
- Numerical analysis (for example, *diff* and *quad*)
- Data analysis (for example, *mean* and *sort*)
- Graphics (for example, *bar* and *polar*)

Complex Numbers

Many engineering problems involve finding the roots of polynomial functions. Such functions often arise when designing or analyzing oscillating systems, such as electronic circuits, springs, shock absorbers, and other mechanical structures that are submitted to shocks. The roots of these polynomials are often complex. Therefore, all MATLAB operations and functions are designed to easily handle complex numbers. You may use either i or j to indicate the imaginary part of any complex number; by default, MATLAB uses i. For example, you can define a variable

```
» a=1+2*i
```

Complex matrices can be defined in the same way. A 2×2 complex matrix can be defined so that

```
» M=[1+2.1*i   2.1+i;   1.1-3.9*i   -0.8-0.2i]
```

When a computed result is complex, MATLAB generates output like the following:

```
» sqrt(-43)

ans=
   0+6.5574i
```

MATLAB includes a few built-in functions that work specifically with complex numbers. For example, *real*(*c*) returns the real part of a complex number *c*, *imag*(*c*) returns the imaginary part of *c*, and *conj*(*c*) returns the complex conjugate of *c*.

EXAMPLE 4-1

Finding the Complex Roots of a Quadratic Function

Determine the roots of the quadratic polynomial $y(x) = 4x^2 - 2x + 1$.

SOLUTION

As you solve the example problems of this chapter, remember to press ENTER at the end of each command line and to use the *clear* command as needed to clear out all past variables before starting each new problem. To start this problem, first define the quadratic coefficients using the form $y(x) = ax^2 + bx + c$.

```
» a=4; b=-2; c=1;
```

Then use the quadratic formula to determine the two roots.

```
» R1=(-b+sqrt(b^2-4*a*c))/(2*a)

R1=
   0.2500+0.4330i

» R2=(-b-sqrt(b^2-4*a*c))/(2*a)

R2=
   0.2500-0.4330i
```

Since our focus in this example was on dealing with complex numbers, we used the well-known quadratic formula to find the roots of the equation. Later in this section, you will be shown how to more easily determine the roots of polynomials of any degree, including quadratic, using the MATLAB *roots* function. The *fzero* function, which returns the roots of nonlinear functions, is covered in the following subsection.

Try It

Use the quadratic formula to determine the roots of the quadratic polynomial $y(x) = 8x^2 - x + 2$.

Minimums and Zeros of Nonlinear Equations

Some of MATLAB's matrix functions take another function as one of their arguments and are therefore called *function functions*. Examples of function functions are *roots*, which returns the roots of a nonlinear function; *diff*, which returns the derivative of a function; and *quad*, which returns the integral of a function. We will cover nonlinear equation solution in this subsection; numerical differentiation and integration we will cover later in this section.

The functions on which function functions operate can either be built in or user defined in M-files. A typical form of a function function is

```
fmin('funcname', lowerlimit, upperlimit)
```

In this case, the *fmin* function finds a local minimum of the *funcname* function that is defined in the M-file, FUNCNAME.M. The arguments, *lowerlimit* and *upperlimit*, set the limits between which MATLAB searches for the minimum. Other commonly used function functions are *fzero*('*funcname*', *X0*), which finds a zero of the function *funcname* near the point *X0*, and *fplot*('*funcname*', *limits*, *numpoints*), which plots the function *funcname* between the limits defined by *limits*. The *fplot* upper and lower limits are defined by a matrix *limits* that has the form [*lowerlimit upperlimit*].

EXAMPLE 4-2

Finding the Local Minimum and Root of a Nonlinear Function

Define the function $f(x)=\sin(x)/x$ in an M-file. Remember to place a period before the / in the function so that an element-by-element division is performed. Remember also to add a trailing semicolon at the ends of the commands so that the command results are not echoed. Plot $f(x)$, and determine the minimum and zero of $f(x)$ between $x = 0$ and $x = 5$.

SOLUTION

First, open the editor/debugger window by selecting M-file from the New option under the File menu. Type the following in the editor/debugger window:

```
function y=f(x)
% This function computes f(x)=sin(x)/x on an element by
element basis
y=sin(x)./x;
```

Select Save As under the File menu in the editor/debugger window, and save the file as F.M. Close the editor/debugger window by selecting Exit Editor/Debugger under the File menu. Now, to plot $f(x)$ from $x = -20$ to $+20$, use the function plotting function *fplot*.

```
» fplot('f',[-20 20])
» title('sin(x)/x')
```

The screen should now display

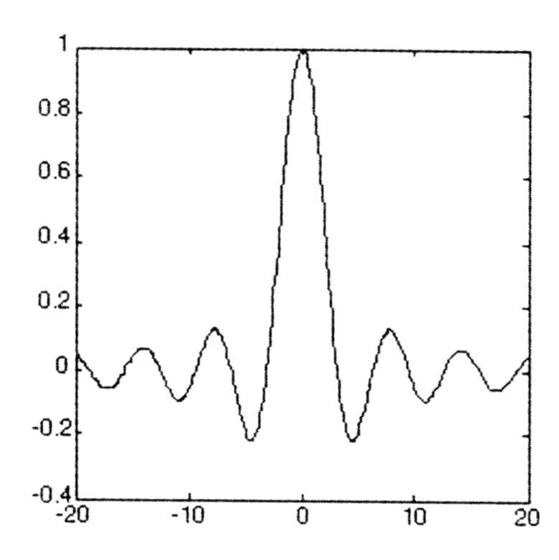

Now, determine the minimum and the zero between $x = 0$ and $x = 5$.

```
» fmin('f',0,5)

ans=
   4.4934

» f(ans)

ans=
   -0.2172
```

The *fzero* function requires two parameters: the name of the function and a "guess" about where the function has a value of zero. We can see from the plot of our function $f(x)$ that f has a value near $x = 2.5$, so we type

```
» fzero('f',2.5)

ans=
   3.1416
```

The minimum between $x = 0$ and $x = 5$ occurs at $x = 4.4934$, where $y = -0.2172$. The zero occurs at $x = 3.1416$. You can see from the plot of $f(x)$ that these are reasonable answers.

Try It Define the function $g(x) = x^3 - 20x^2$ in an M-file. Plot $g(x)$ from $x = -20$ to $+20$ and determine the minimum and zero of $g(x)$ between $x = 0$ and $x = 10$.

Solving Sets of Simultaneous Equations

Consider the following set of equations, involving three unknowns:

$$
\begin{array}{rrrcr}
x_1 & +x_2 & -x_3 & = & 5 \\
-2x_1 & -2x_2 & +x_3 & = & -5 \\
3x_1 & +2x_2 & -3x_3 & = & 0
\end{array}
$$

There are several ways to solve a system of simultaneous equations using MATLAB. One way is to use Cramer's rule, which involves the use of determinants. An easier way is to use matrix division. For example, the preceding equations can be written in the following matrix form:

$$\mathbf{CX} = \mathbf{B}$$

where

$$
\mathbf{C} = \begin{array}{rrr} 1 & 1 & -1 \\ -2 & -2 & 1 \\ 3 & 2 & -3 \end{array} \qquad \mathbf{X} = \begin{array}{c} x_1 \\ x_2 \\ x_3 \end{array} \qquad \mathbf{B} = \begin{array}{r} 5 \\ -5 \\ 0 \end{array}
$$

Then, the values of the unknowns we require are **X** = **B**/**C**.

The matrix multiplication, **CX** = **B**, can also be expressed as **X'C'** = **B'**, where **X'**, **C'**, and **B'** are the transposes of **X**, **C**, and **B**, respectively. Then, **X'** = **B'**/**C'**, the transpose of the solution we seek. MATLAB can find **X** more easily and more quickly in this way than it can in the more straightforward way of solving **X** = **B**/**C**. Example 4-3 demonstrates the process.

EXAMPLE 4-3

Solving a Set of Simultaneous Equations

Solve the following set of simultaneous equations:

$$
\begin{array}{rrrcr}
x_1 & +x_2 & -x_3 & = & 5 \\
-2x_1 & -2x_2 & +x_3 & = & -5 \\
3x_1 & +2x_2 & -3x_3 & = & 0
\end{array}
$$

SOLUTION

We start by defining the coefficient matrix **C** and the matrix **B**.

```
» C=[1  1  -1;  -2  -2  1;  3  2  -3];
» B=[5;  -5;  0];
```

Now, we perform the matrix division to find the transpose of the **X** vector.

```
» XTrans=B'/C'

XTrans=
   -15.0000   15.0000   -5.0000
```

Therefore, the solution of the set of equations is $x_1 = -15$, $x_2 = 15$, and $x_3 = -5$.

Try It

Solve the following set of simultaneous equations:

$$\begin{aligned} -x_1 \quad +2x_2 \quad -x_3 &= 15 \\ -x_1 \quad -x_2 \quad +x_3 &= -5 \\ 3x_1 \quad -2x_2 \quad -2x_3 &= 10 \end{aligned}$$

Roots of a Polynomial

MATLAB can easily determine the roots of an arbitrary polynomial using the *roots(C)* function. If the coefficients are stored in vector **C**, the roots are returned by *roots(C)*. For example, to determine the roots of the simple quadratic function $f(x) = 2x^2 + x + 2$,

```
» C=[2 1 2];
» roots(C)

ans=
   -0.2500+0.9682i
   -0.2500-0.9682i
```

EXAMPLE 4-4

Finding the Roots of a Polynomial

Determine the roots of the polynomial function $f(x) = 2x^4 - 3x^2 - 5$. Check to make sure the roots are correct.

SOLUTION

First define the coefficient matrix.

```
» coeff=[2  0  -3  0  -5];
```

Note that the coefficients of x^3 and x^1 are zero.

Now, find the roots; there will be four roots since the polynomial is of degree four.

```
» Roots=roots(coeff)

Roots=
    1.5811
   -1.5811
      0+1.0000i
      0-1.0000i
```

Note that two of the roots are real and two are imaginary.

You could check the accuracy of these roots by using the editor/debugger to create the user-defined function $f(x) = 2x^4 - 3x^2 - 5$. Name the resulting file CHECK.M. Then, in the MATLAB command window, type

```
» check(Roots(1))
» check(Roots(2))
» check(Roots(3))
» check(Roots(4))
```

The answers may not be exactly zero because of rounding errors; however, they should be on the order of 10^{-15}.

Try It Determine the roots of the polynomial function $f(x) = 3x^3 - 2x^2$. Then, create an M-file definition of this function, and check the roots to confirm that they are correct.

Linear and Polynomial Regression

Linear regression is a mathematical process that determines the best linear fit through a set of data points. It is also referred to as *least squares fit* because the line that best fits the data is that line that minimizes the sum of the squared distances of the data points from the line. *Polynomial regression* is a process of finding that polynomial function that best fits a set of data points. The function that results from the regression process, whether linear or of higher degree, is used to estimate data points that lie between the known data points, a process called *interpolation.*

The MATLAB function *polyfit(x,y,n)* uses polynomial regression to return the coefficients of the polynomial that best fits the data points whose X and Y coordinates are, respectively, in the vectors **x** and **y**. The degree of the polynomial is *n*. Therefore, to perform a linear regression, you would use $n = 1$. The number of coefficients returned is $n + 1$. Example 4-5 demonstrates the use of *polyfit* for performing a linear regression and a third degree polynomial regression.

EXAMPLE 4-5

Polynomial Regression

The following data represent voltage samples, in volts, taken in a circuit once every millisecond for 8 milliseconds:

time =	1	2	3	4	5	6	7	8
voltage =	1.1	2.0	3.4	3.5	3.2	3.4	5.0	7.0

Perform both a linear regression and a third degree polynomial regression on the data points.

SOLUTION

You start by defining the **time** vector, in which the time equals 1 to 8 milliseconds.

```
» time=1:8;
```

Next, define **volts**. Since the voltage values are not evenly spaced, you can't use the shorthand method, as you did for the **time** data, to define them. Instead, you must define each data point, as follows:

```
» volts=[1.1 2.0 3.4 3.5 3.2 3.4 5.0 7.0];
```

Now, plot the data points.

```
» plot(time,volts,'o')
```

The 'o' in the *plot* function call tells the function to show lowercase o's at each data point. The default action, produced by the command *plot(time,volts)*, is to connect the data points with straight lines. We didn't use the default because we want to find the "best fit" through the data, not just connect the data points.

A figure window is created, which should show

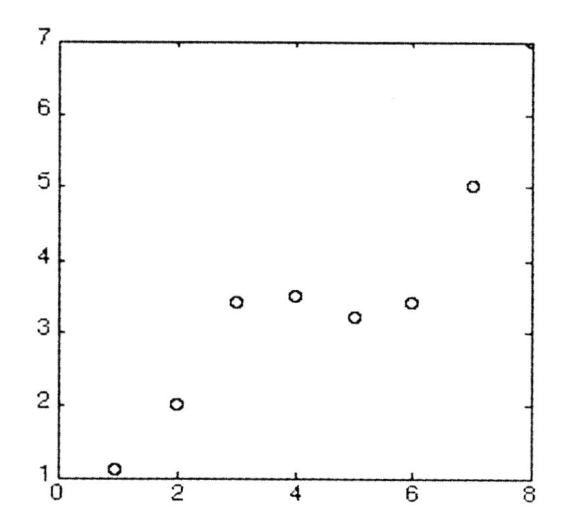

Now, hold this plot in the figure window while you plot the linear and third degree polynomial in the same window.

```
» hold on
```

Now, determine the coefficients of the function that plots a best linear fit through the data points.

```
» coeff=polyfit(time,volts,1)

coeff=

  0.6667   0.5750
```

Use these coefficients to compute eight points on the straight line, corresponding to *time* = 1 to *time* = 8, the values in the vector **time**.

```
» line=coeff(1).*time+coeff(2);
```

Plot this straight line through the data points already plotted.

```
» plot(time,line)
```

The figure window should now show

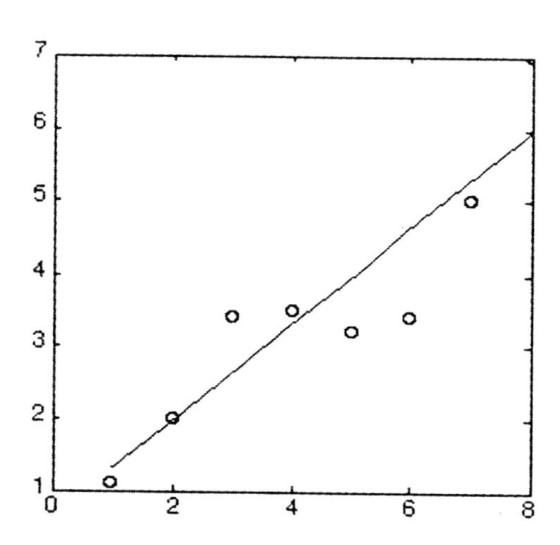

Interpolate to find the voltage at *time* = 2.3 milliseconds.

```
» voltage=coeff(1).*2.3+coeff(2)

voltage=
  2.1083
```

Now, determine the coefficients of the third degree polynomial that is a best fit for the plotted data point.

```
» coeff=polyfit(time,volts,3)

coeff=
  0.0687   -0.8689   3.6790   -1.9500
```

Use these coefficients to compute data points that correspond to the *time* data values.

```
» poly=coeff(1).*time.^3+coeff(2).*time.^2+...
  coeff(3).*time+coeff(4);
```

Plot the third degree polynomial.

```
» plot(time,poly)
```

Finally, the figure window should show

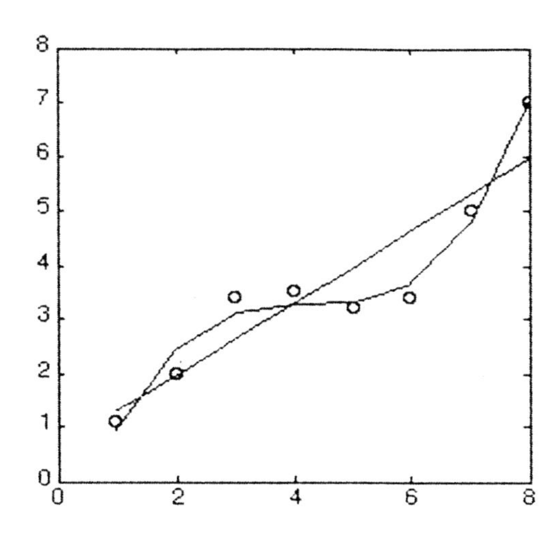

You can see from the plot that the third degree polynomial fits the data better than does the linear equation. To confirm that it provides a better fit, compute $t = 2.3$ seconds using the third degree polynomial, as follows:

```
» voltage=coeff(1).*2.3.^3+coeff(2).*2.3.^2+...
  coeff(3).*2.3+coeff(4)

voltage=
  2.7508
```

Compare this result to the 2.1083 volts computed from the linear regression equation. Looking at the original data points in the graph, it seems likely that 2.7508 volts is closer to the actual value, had it been measured.

Of course, before printing or saving a plot such as this, you would want to add appropriate labels. For example, the following would do:

```
» xlabel('Time'); ylabel('Volts'); title('Voltage Samples')
```

Try It

The following represents pressure samples, in psi, taken in a fuel line once every second for 8 seconds:

time =	1	2	3	4	5	6	7	8
pressure =	51	55	58	52	49	53	57	52

Perform both polynomial regressions on the data points to determine the degree, less than five, that best fits the data.

Numerical Differentiation

Engineers often use differentiation to determine the rate of change of a process when changes are known, and integration to determine the changes in a process when the rate of change is known. Graphically, the

derivative of a function $f(x)$ at a point x is the slope of the curve at $x=a$ that is defined by the function. By definition, the derivative of a function $f(x)$ is a function $f'(x)$ that is equal to the rate of change of $f(x)$ with respect to x. The notation used to represent this value is

$$\frac{df(x)}{dx}$$

The MATLAB *diff(f)* function computes differences between adjacent values in a vector, returning a vector that has one less value. As with other matrix functions, *diff* will operate on the columns of a matrix passed to it, returning a matrix in which each column has one less value. An approximate derivative $df(x)/dx$ can be computed using *diff* by dividing a change in x into a corresponding change in $f(x)$.

```
» derivative=diff(f(x))./diff(x)
```

The accuracy of a derivative generated by *diff* depends on the resolution with which the differentiated function is computed. In Example 4-6, $f'(x)$ is computed over the interval $x = 0$ to 5 with resolutions of 0.1, 0.01, and 0.001. In each case, the derivative at $x = 2.5$ is computed, and the results are compared.

EXAMPLE 4-6

Performing Numerical Differentiation

Determine the linear acceleration of an object whose speed is defined by $s(t) = t^3 - 2t^2 + 2$ meters/second, where t is in seconds, over the interval $t = 0$ to 5. Then determine the specific acceleration at $t = 2.5$ seconds.

SOLUTION

You will compute the derivative function at three different resolutions to show the effect of increasing the resolution. Begin with a speed resolution of 0.1 second.

```
» t=0:0.1:5;
```

Compute the speed at each of the 51 time values.

```
» s=t.^3-2*t.^2+2;
```

Compute the acceleration at each of the 51 speed values by computing the approximate derivative.

```
» ds=diff(s)./diff(t);
```

To determine the acceleration (the derivative of the speed with respect to time) at $t = 2.5$ seconds, you must note that you want the 26th element in the vector **ds**. This is because the indices of **ds** begin at 1, whereas t values begin at 0, and there are ten elements per second. Since we want the element corresponding to $t = 2.5$ seconds, the index we need is $1 + 2.5 \cdot 10 = 26$.

```
» ds(26)

ans=

  9.3100
```

That is, 9.31 meters/second^2. Now, start over with a resolution of 0.01 second.

```
» t=0:0.01:5;
» s=t.^3-2*t.^2+2;
» ds=diff(s)./diff(t);
```

This time, we want the element with the index $1 + 2.5 \cdot 100 = 251$.

```
» ds(251)

ans=

  8.8051
```

Repeat the process one more time using a resolution of 0.001 second. This time, we want the element with the index $1 + 2.5 \cdot 1000 = 2501$.

```
» t=0:0.001:5;
» s=t.^3-2*t.^2+2;
» ds=diff(s)./diff(t);
» ds(2501)

ans=

  8.7555
```

The answer is decreasing with each increase in resolution, converging toward the correct answer. In order to test your work, you can analytically integrate $s(t)$ to obtain $s'(t) = 3s^2 - 4s$. Plugging $t = 2.5$ seconds into this equation yields an exact answer of 8.75 m/s^2. The point is that, since the MATLAB *diff* function returns only an approximate derivative, you must use the resolution required to achieve the accuracy you desire. The higher the resolution, the smaller should be the range over which you compute the derivative. If you combine high resolution with a large range, very large vectors will be created that may cause out-of-memory errors. In the problems at the end of the chapter, you are asked to write a program that returns a derivative with a specific resolution.

Try It The current, in amps, through a capacitor in an electronic circuit is related to the derivative with respect to time of the voltage across that capacitor in the following way:

$$i = C\,dv/dt$$

where C is the capacitance of the capacitor in Farads.

Determine the current at $t = 0.5$ second through a 2-Farad capacitor, the voltage across which is $v(t) = 5e^{0.5t}$, where t is in seconds.

Numerical Integration

Graphically, the integral of a function $f(x)$ is the area under the curve defined by the function. This area is usually computed over a finite interval although it may be computed between negative and positive infinity for those functions for which such an integral is not itself infinity. By definition, then, the integral of a function $f(x)$ over an interval $[ab]$ is the area under the curve defined by $f(x)$ between a and b. The notation used to represent this value is

$$\int_a^b f(x)\,dx$$

There are a number of algorithms by which an integral can be numerically calculated. One method involves dividing up the area under the curve into trapezoids and then adding the areas of the trapezoids. As in the case of the method for computing derivatives, the accuracy of the result is determined by the resolution with which the trapezoids are created. The smaller the trapezoids, the greater the accuracy.

Although you could use the trapezoid method to compute integrals using MATLAB, the method involves considerable mathematics. A more convenient way to compute integrals is to use the MATLAB *quad('funcname',a,b)* function. The term *quad* is short for "quadrature," an algorithm that uses an adaptive form of Simpson's rule to calculate integrals. The input arguments are *'funcname'*, the name of a built-in or user-defined function, and *a* and *b*, the starting and ending limits of the interval over which the integral is taken.

EXAMPLE 4-7

Performing Numerical Integration

Use MATLAB's *quad* function to compute the integrals of the functions $f(x) = \sin(x)$ and $g(x) = \sin(x)/x$ over the interval $x = \pi/2$ to π.

SOLUTION

The *sin* function is built in, making this integral easy.

```
» quad('sin',pi/2,pi)

ans=
  1.0000
```

You may have defined the function $\sin(x)/x$ in an earlier example and saved it as F.M. To check, enter the command *type f.m.* If the function has not been defined, you must define it as follows.

Open the editor/debugger window and type (remember the period before the /)

```
function y=f(x)
% This function computes sin(x)/x on an element by element
basis.
y=sin(x)./x;
```

Save the function as F.M and close the editor/debugger window.
Now, compute the desired integral, naming the function you just created.

```
» quad('f',pi/2,pi)

ans=

  0.4812
```

The value 0.4812 is the area under the curve defined by $\sin(x)/x$ over the interval $x = \pi/2$ to π.

Try It

Compute the integrals of the functions $f(x) = e^x$ and $g(x) = 5e^x\sin(x)$ over the interval $x = 0$ to π.

4-2 DATA ANALYSIS

Engineers must often analyze data. These data may be collected when making measurements that will be used in the design of an engineering system, such as bridges or electronic circuits. The data may also be collected as an engineering system is tested to make sure it meets its original specifications. MATLAB includes numerous functions that aid in data analysis. In this section, we will describe only the most commonly used data analysis functions.

Random Numbers

Engineers often need to generate random numbers that are then used as input to test a hardware model or software algorithm. The *rand* function in MATLAB is used to generate random numbers. There are two basic forms of the function.

- *rand*(n) returns an $n \times n$ matrix of random numbers.
- *rand*(m,n) returns an $m \times n$ matrix of random numbers.

Both of these functions return random numbers uniformly distributed in the interval (0,1). As a simple example of a use of random numbers, you could simulate a coin toss with the following user-defined function:

```
function cointoss
% No arguments are passed to or from this function
if rand(1)<0.5
  disp('heads')
else
  disp('tails')
end
```

Once the function has been saved in an M-file named COINTOSS.M, you would use it in the MATLAB command window so that

```
» cointoss

tails

» cointoss

heads
```

Although uniform distributions of random numbers are useful in many applications, as an engineer you may also work with data that is normally distributed. To generate normally distributed random numbers (also called Gaussian numbers), you must use either *randn*(n) or *randn*(m,n). These two functions return random numbers that have a mean of 0 and a variance of 1.

Note that each of the *rand* functions will return the same sequence of random numbers each time you run MATLAB. For example, *rand*(2) will always generate

```
» rand(2)

ans=
   0.2190   0.6789
   0.0470   0.6793
```

the first time you use it. The reason for this is that MATLAB uses a seed number to start its random number generator, and MATLAB always starts with the same seed. You can change the seed with the command *rand*('*seed*',n), where n is the new seed number.

As we stated, the *rand* function returns random numbers and matrices whose elements are distributed in the interval (0,1). But rarely do engineers desire random numbers between 0 and 1. Instead, they must be able to generate random numbers that have a range similar to the real-world data that will be input into a hardware or software system. And in many cases, they may want to generate integers rather than real numbers to simulate real-world data.

Example 4-8 shows how to generate random numbers of any range and how to convert random real numbers into integers using the *round*(x) function, which rounds x to the nearest integer. (MATLAB offers three other rounding functions: *fix*(x), which rounds x toward zero; *floor*(x), which rounds x toward $-\infty$; and *ceil*(x), which rounds x toward $+\infty$.)

EXAMPLE 4-8 Generating Random Numbers

Generate a vector of seven random real numbers that range uniformly between -10 and $+10$. Then generate a 3×3 matrix of random integers that range uniformly from 0 to 10.

SOLUTION

Since you want a range of 20 (-10 to +10) instead of the range of 1 generated by *rand*, you multiply the output of *rand* by 20. Since the range begins at -10 instead of 0, as generated by *rand*, you add -10 to the result in the following way:

```
» R=rand(1,7)*20-10

R =

  9.0026  -5.3772  2.1369  -0.2804  7.8260  5.2419  -0.8706
```

Since these are random numbers, your values may differ.

To solve the second problem, use the *round* function to convert real numbers to integers:

```
» I=round(rand(3)*10)

I =

  0   6   7
  8   8   2
  4   9   4
```

Try It Generate a 4 × 4 matrix of random real numbers that range uniformly between -10 and -5. Then generate a vector of ten random integers that range uniformly from -10 to -5.

Maximum, Minimum, and Sorting

The collection and analysis of data is often a part of the design process. Data analysis is often as simple as finding the smallest or largest element in the data.

Using MATLAB, determining the minimum or maximum of a vector is as easy as

```
» v=[-2 3 4 -1 0 5];
» min(v)

ans=
   -2

» max(v)

ans=
   5
```

When the argument is a matrix, the *min* and *max* functions return the row vector that contains the minimum or maximum element from each column.

```
» M=[-2 3 4 -1; -4 0 3 7; 0 4 6 -9]

M=

  -2   3   4  -1
  -4   0   3   7
   0   4   6  -9

» min(M)

ans=

  -4   0   3  -9

» max(M)

ans=

   0   4   6   7
```

The *sort(x)* function is used to sort vectors and the columns of matrices. In both cases, the sorting is performed in ascending order.

```
» V=[-2 3 4 -1 0 5];
» Vsorted=sort(V)

Vsorted=

  -2   -1   0   3   4   5

» M=[-2 3 4 -1; -4 0 3 7; 0 4 6 -9];
» Msorted=sort(M)

Msorted=

  -4   0   3  -9
  -2   3   4  -1
   0   4   6   7
```

EXAMPLE 4-9

Sorting the Elements of a Matrix

Create a 3×3 matrix of random real numbers that uniformly range from -100 to 0. Then sort the rows.

SOLUTION

First, create the matrix of random numbers.

```
» M=rand(3)*100-100

M=
   -78.1041   -32.0704   -48.0584
   -95.2955    -6.5307   -16.9035
   -32.1135   -61.6498   -96.5428
```

(Your matrix may differ from **M** shown here.)

Then sort the matrix's transpose, and take the transpose of the result. In this way, the rows are sorted rather than the columns.

```
» M=sort(M');
» M=M'

M=
   -78.1041   -48.0584   -32.0704
   -95.2955   -16.9035    -6.5307
   -96.5428   -61.6498   -32.1135
```

Try It

Create a 4 × 4 matrix of random integers that range uniformly from -25 to 0. Then sort the rows.

Mean, Median, Standard Deviation, Sums, and Products

The following data analysis functions work in the same way as do the *min* and *max* functions. Each works on vectors as a whole or on the columns of a matrix.

mean(x)	Returns the mean of the vector **X** or the means of the columns of matrix **x**
median(x)	Returns the median of the vector **X** or the medians of the columns of matrix **x**
std(x)	Returns the standard deviation of the vector **X** or the standard deviations of the columns of matrix **x**
sum(x)	Returns the sum of the vector **X** or the sums of the columns of matrix **x**
prod(x)	Returns the product of the vector **X** or the products of the columns of matrix **x**

The following is an example of using *sum* and *prod* on a matrix:

```
» M=rand(3)*200-100

M=
   -56.2082    35.8593     3.8833
   -90.5911    86.9386    66.1931
    35.7729   -23.2996   -93.0856

» S=sum(M)

S=
  -111.0263    99.4983   -23.0092

» P=prod(M)

P=
  1.0e+005*
  1.8215   -0.7264   -0.2393
```

EXAMPLE 4-10

Determining Mean and Standard Deviation

Create a 3 × 3 matrix of random integers that range from -3 to +3. Then determine the mean and standard deviation of each column in the matrix.

SOLUTION

```
» M=round(rand(3)*6-3)

M=
  -3   -3   0
   0   -1   1
   1   -3   1

» mean(M)

ans =
  -0.6667   -2.3333   0.6667

» std(M)

ans =
  2.0817    1.1547    0.5774
```

Try It

Create a 4 × 4 matrix of random real numbers that range from -2 to +2. Then determine the sum, mean, and median of each column in the matrix.

Histograms

A histogram is a bar graph that plots the number of times each number in a set of numbers appears in that set. The MATLAB function *hist(x)* generates and plots a histogram of the numbers in a vector **x**. By default, *hist* separates the numbers in **x** into ten equally spaced bins; therefore, if the numbers range from 0 to 10, *hist* would plot a histogram in which there are ten bars, indicating the number of numbers with values from 0 to 1, 1+ to 2, 2+ to 3, 3+ to 4, 4+ to 5, 5+ to 6, 6+ to 7, 7+ to 8, 8+ to 9, 9+ to 10. If you desire a different number of bins, use the form *hist(x,n)*, in which *n* is the number of bins desired.

To plot a 25-bar histogram of the numbers in vector **x**, you would enter

```
» hist(x,25)
```

To plot a 25-bar histogram of the second row of matrix **X**, you would enter

```
» hist(X(2,:),25)
```

Another form of the *hist* function is *hist(x,y)* where **y** specifies the bins into which the elements of **x** are placed. Either of the following would create a histogram in which the elements of **x** are placed in the 20 bins -10 to -8, -8 to -6, ..., 8 to 10:

```
» i=-10:2:10
» hist(x,i)
```

or simply

```
» hist(x,-10:2:10)
```

EXAMPLE 4-11

Plotting a Histogram

Assume that, in the following matrix, the first column represents the identification numbers of a series of oil wells, the second column lists each well's average daily production, and the third column lists the lease number in which each well is located. Plot a histogram that shows the number of wells that produce 0 to 10 barrels of oil per day (bopd), 10 to 20 bopd, 20 to 30 bopd, and 30 to 40 bopd.

```
data = 101   3  23
       103  23  23
       104  16  23
       107  18  23
       115  45  26
       117  32  26
       121  29  31
       122  39  31
       129   7  31
```

SOLUTION

First, enter the matrix.

```
» data=[ 101 3 23;  103 23 23;  104 16 23;  107 18 23;...
115 45 26;  117 32 26;  121 29 31;  122 39 31;  129 7 31];
```

Now, since you desire the groupings, 0 to 10, ..., 30 to 40, plot the histogram as follows:

```
» hist(data(:,2),0:10:40)
```

The figure window should show

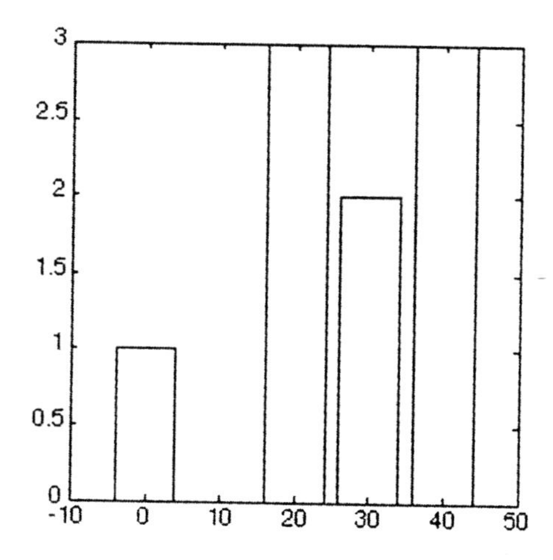

Try It Create a 12-bar histogram of a 100-element vector of normally distributed random integers that range from 0 to 100.

4-3 PLOTS

By graphing data, an engineer can often spot, at a glance, patterns or anomalous data that would require hours of programming to detect mathematically without a visual aid. As we stated earlier, MATLAB excels as just such a visual aid. Throughout this and previous chapters, you have been using the X-Y plotting capabilities of MATLAB in order to see the data with which you were dealing. In this section, you will learn the ease with which MATLAB can also plot bar graphs, as well as polar, mesh, and contour plots.

Bar Graphs

Although a bar graph looks similar to a histogram, they are not the same; the height of each bar in a histogram represents the number of numbers in a bin. The height of each bar in a bar graph represents the value of an element in an array. For example, the command *bar*(*a*) will plot a vertical

bar graph in which the number of bars is the same as the number of elements in the vector **a**, and the height of each bar is a measure of the value of the element it represents. As with other plots, the color of the bars can be specified using a third argument in quotes. For example, *bar(a,'g')* plots green bars. (See Section 2-2 for a list of other colors and line types that can be specified.) As with all graphs, a title and X-axis and Y-axis labels can be added.

A second form of this command, *bar(x,y)*, creates a bar graph of the elements of the vector **y** at the locations specified in the vector **x**. The vector **x** must contain equally spaced ascending values. Example 4-12 illustrates the difference between using *bar(x)* and *bar(x,y)*.

EXAMPLE 4-12

Creating a Bar Graph

Create a bar graph of $f(w) = \sin(w)$ for $w = 0$ to 2π in increments of $\pi/15$. Use both forms of the *bar* function and compare the difference. Place on top of the second bar graph a dashed line plot of $g(w) = \sin(w - 22.5°)$.

SOLUTION

First, define the following 31 elements in vector **w**:

```
» w=0:pi/15:2*pi;
```

Create a vector **f** that contains the *sine* of the 31 elements in **w**:

```
» f=sin(w);
```

Create the bar graph.

```
» bar(f)
```

The figure window should now show

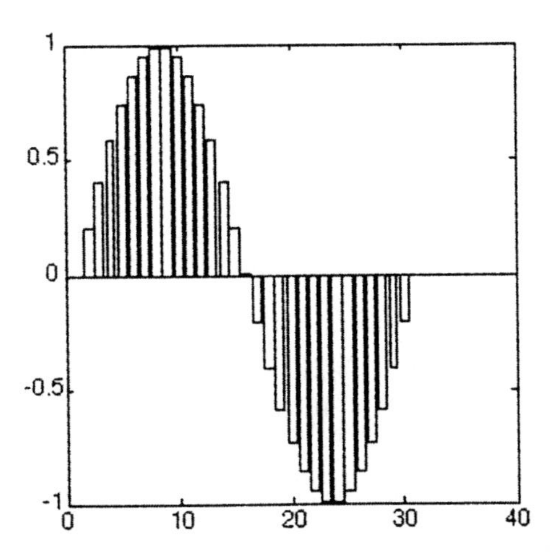

Notice that, in this graph, the bars are placed along the X-axis from 1 to 31. The X-axis values represent the 31 indices of the **f** vector elements, the values of which the height of the bars represent.

Now create a second bar graph that plots the elements in vector **f** at the points specified by the elements in vector **w**.

```
» bar(w,f)
```

Hold this bar graph in the figure window.

```
» hold on
```

Now, create the X-Y plot, delaying the *sine* curve by 22.5°, or $\pi/8$ radians.

```
» h=sin(w-pi/8);
» plot(w,h,'--')
```

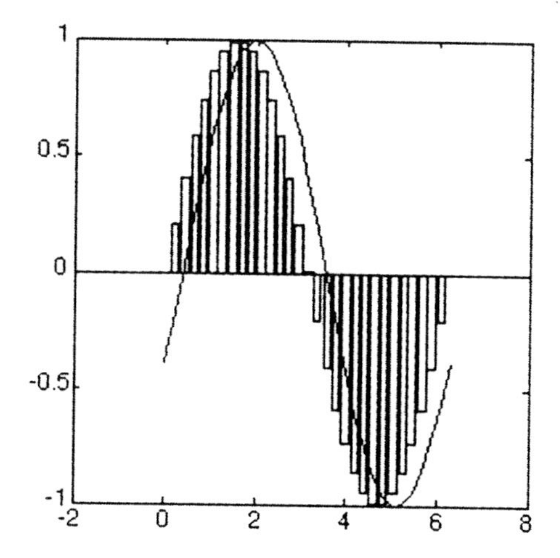

Notice that, in this graph, the bars are placed at $x = 0$ to 2π, as specified by the **w** vector in *bar(w,f)*, rather than 1 to 31, as in the first bar graph.

Try It

Create a bar graph of e^x for $x = 0$ to 5 in increments of 0.5. Use both forms of the *bar* function, using two different colors, and compare the difference.

Polar Plots

In an X-Y plot, each point has an X coordinate and a Y coordinate. Points in a polar plot are also represented by two values; however, these two values are an angle and a magnitude. Generally, for each angle θ from 0 to 2π radians, or 0° to 360°, a point is plotted a positive distance, a magnitude, from the origin of the graph. However, θ can range beyond 2π, and magnitudes can be negative.

You use polar plots when you want to visualize data values that represent angles and magnitudes. For example, polar plots are ideal for visualizing a flux field around a magnet or light intensity around a light source. Polar plots are also useful for plotting complex values since they can be viewed as vectors with angle and magnitude.

The MATLAB *polar(theta,r)* function generates polar plots using the elements in **theta** as angles (in radians) and the elements in **r** as correspond-

ing magnitudes. For example, the following plots five points, each with a magnitude (or radius) of 1, at 0°, 90°, 180°, 270°, and 360°:

```
» r=[1 1 1 1 1];
» x=[0  pi/2  pi  3*pi/2  2*pi];
» polar(x,r)
```

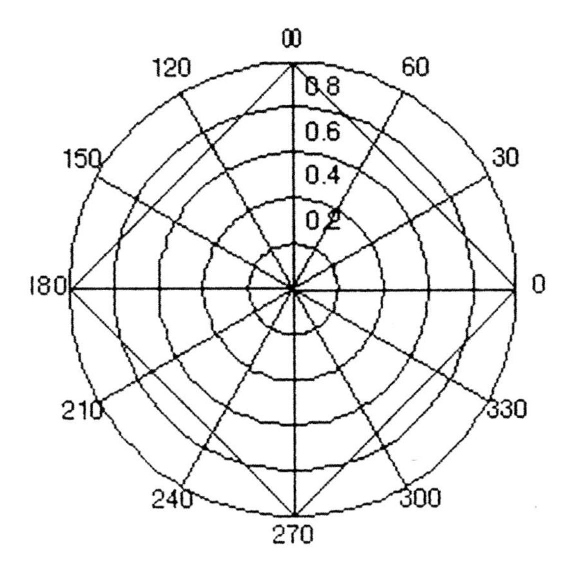

By default, as with X-Y plots, MATLAB connects the points with a solid straight line. Alternative line types and colors can be specified, as with previously described plots, using a third argument that is enclosed in quotes.

EXAMPLE 4-13

Creating a Polar Plot

Create a polar plot of a decreasing radius curve that starts at a magnitude of 1 at an angle of 0° and ends at a magnitude of 0 at an angle of 360°.

SOLUTION

The first issue is to choose the resolution with which you want to draw this curve. A resolution that is too low results in a curve that shows discontinuities, as in the previously discussed five-point plot. A resolution that is too high wastes memory space and takes an unnecessarily long time to compute points and create the plot. Clearly, plotting one point per radian is too low since there are only 6.28 radians in 360°. You might consider plotting 360 points, one per degree; however, 100 points is usually sufficient.

```
» angle=0:2*pi/100:2*pi;
```

The next command creates a vector **magnitude**, the same size as is **angle**, in which the values of the elements decrease linearly from 1 to 0 as *angle* increases linearly from 0 to 2π.

```
» magnitude=1-angle/(2*pi);
```

If the plot is currently held, turn off hold with the *hold off* command. Now, create the polar plot and give it a title.

```
» polar(angle,magnitude)
» title('Polar Plot')
```

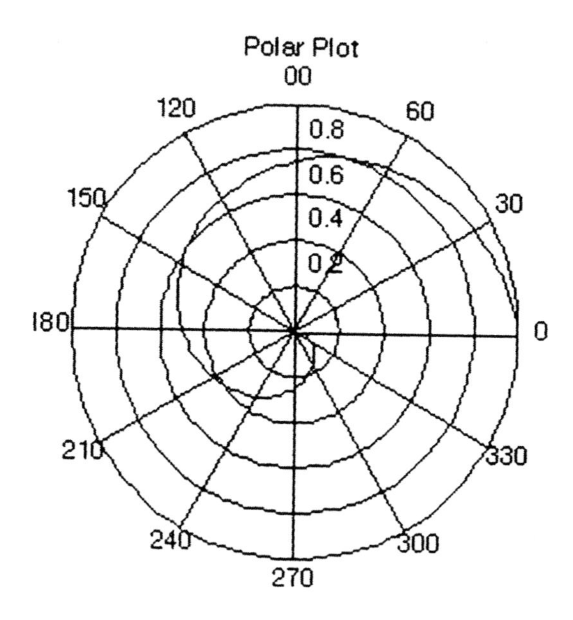

Try It Create a polar plot of an increasing radius curve that starts at a magnitude of 0 at an angle of 0° and ends at a magnitude of 4 at an angle of 720°.

Contour Plots

Contour maps are often used by civil engineers to show the elevations, above sea level, over the surface of a plot of land. In this way, high and low points can be easily seen, aiding engineers in, for example, the planning of the construction of a highway or building. A contour plot can also be used to show the pressure over a surface. Typically, points of equal value on contour maps are connected by straight lines.

The MATLAB *contour(X)* function plots a contour map of the elements in a matrix **X**. The lower left-hand corner of the plot corresponds to $X_{1,1}$. The *contour* function connects points of equal value with straight lines; however, the number of contour lines and their values are chosen automatically by the function. You can use the alternative form *contour (X,n)* to force *contour* to plot exactly *n* contour lines.

By default, the contour lines are multicolored. If you prefer that MATLAB use a grayscale, you can use the *colormap(gray)* command. A grayscale plot produces a better hardcopy if you do not have a color printer. You can print a hardcopy of any MATLAB plot by selecting Print under the File menu of the figure window.

By default, *contour* labels the axes with the indices of the matrix elements. You can label the axes with your own values by using the form *contour(a,b,X)*. The **a** and **b** vectors would contain increasing values with which the X-axis and Y-axis, respectively, are labeled.

EXAMPLE 4-14

Graphing a Contour Plot

Create a contour map of the land elevation data in the following matrix:

```
data = 1 1 1 2 3 2 1 1 1
       1 1 2 3 4 3 2 1 1
       1 1 2 3 4 3 2 1 1
       1 2 3 4 5 4 3 2 1
       1 2 3 4 5 4 3 2 1
       1 3 4 5 6 5 4 3 1
       1 3 5 6 7 6 5 3 1
       1 2 3 4 5 4 3 2 1
       1 1 2 3 4 3 2 1 1
```

SOLUTION

```
» data=[1 1 1 2 3 2 1 1 1;...
1 1 2 3 4 3 2 1 1; 1 1 2 3 4 3 2 1 1;...
1 2 3 4 5 4 3 2 1; 1 2 3 4 5 4 3 2 1;...
1 3 4 5 6 5 4 3 1; 1 3 5 6 7 6 5 3 1;...
1 2 3 4 5 4 3 2 1; 1 1 2 3 4 3 2 1 1];
» contour(data)
```

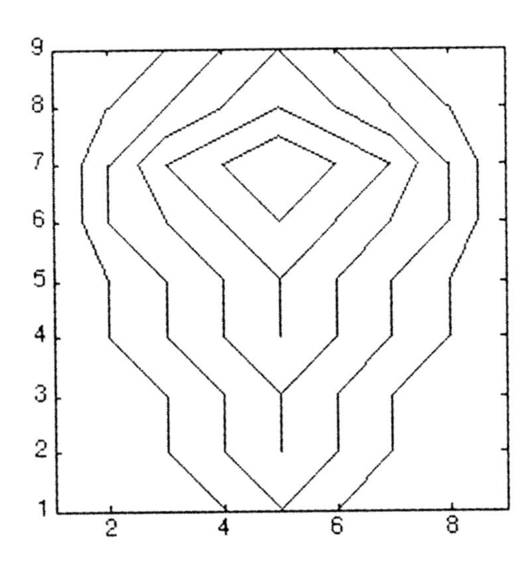

Note that the numbers along each axis are the index values of the plotted matrix elements. Note that point (1,1) is in the lower left-hand corner; therefore, the plot is upside-down relative to the matrix that is plotted. Typically, in a land survey these indices would represent points on the land that are equal distances apart. For example, the plot might be of a piece of land that is 800 feet by 800 feet. If you wanted to specifically label each axis 0 to 800, you would use the command *contour(a,b,data)* after creating the **a** and **b** vectors as follows:

```
» a=[0 100 200 300 400 500 600 700 800];
» b=[0 100 200 300 400 500 600 700 800];
```

The elements in **a** label the X-axis, and the elements in **b** label the Y-axis. The matrix need not be square, but the number of elements in the **a** vector must be the same as the number of rows in **data**, and the number of elements in **b** must be the same as the number of columns in **data**.

Try It

In the same figure window, create contour plots of a 3 × 3 identity matrix and a 3 × 3 magic matrix.

3-D Mesh Plots

The MATLAB *mesh*(*Z*) function plots 3-D surfaces. It takes a matrix as its input argument; the value of each element in the matrix represents a Z-axis value. In Example 4-15, you create a 3-D mesh plot of the same data used to create a contour in the previous example.

EXAMPLE 4-15

Graphing a Mesh Plot

Create a 3-D mesh plot of the land elevation data in the following matrix:

```
data = 1 1 1 2 3 2 1 1 1
       1 1 2 3 4 3 2 1 1
       1 1 2 3 4 3 2 1 1
       1 2 3 4 5 4 3 2 1
       1 2 3 4 5 4 3 2 1
       1 3 4 5 6 5 4 3 1
       1 3 5 6 7 6 5 3 1
       1 2 3 4 5 4 3 2 1
       1 1 2 3 4 3 2 1 1
```

SOLUTION

```
» data=[1 1 1 2 3 2 1 1 1;...
1 1 2 3 4 3 2 1 1; 1 1 2 3 4 3 2 1 1;...
1 2 3 4 5 4 3 2 1; 1 2 3 4 5 4 3 2 1;...
1 3 4 5 6 5 4 3 1; 1 3 5 6 7 6 5 3 1;...
1 2 3 4 5 4 3 2 1; 1 1 2 3 4 3 2 1 1];
» mesh(data)
```

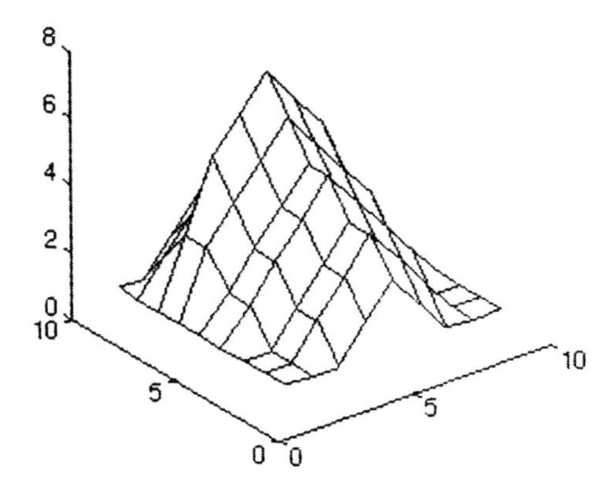

As does the *contour* function, *mesh* labels the X-axis and Y-axis with the indices of the matrix elements. You can label the axes with the *x* and *y* values by using the form *mesh(x,y,Z)*; this form is used in the next example.

3-D plots of two variables, such as $z=f(x,y)$, can be generated with the aid of the *meshgrid(x,y)* function. The *meshgrid* function plots a function of two variables. To use it, you must first generate **X** and **Y** matrices consisting of repeated rows and columns, respectively, over the domain of the function. You then use these matrices to evaluate and graph the function. Example 4-16 demonstrates the process.

EXAMPLE 4-16

Graphing a 3-D Function

Create a 3-D plot of the function $f(x,y) = \sin(x^2 + y^2) - \cos(x^2 + y^2)$ over the range of $x = 0$ to 2 and $y = 0$ to 3 in increments of 0.05.

SOLUTION

First, create the vectors **x** and **y**.

```
» x=0:.05:2;
» y=0:.05:3;
```

Note that **x** contains 41 elements and **y** contains 61 elements. Now, create two 61×41 matrices **X** and **Y** (remember that MATLAB is case sensitive) that are used to create a 61×41 **Z** matrix that contains the *z* values that will be plotted over the x-y plane.

```
» [X Y]=meshgrid(x,y);
» Z=sin(X.^2+Y.^2)-cos(X.^2+Y.^2);
```

Next, use the *colormap* function to produce a black plot (default is multicolored), which is best when producing a black-and-white hardcopy. Omit this command if you want a multicolored plot.

```
» colormap([0 0 0])
```

The *colormap* parameter is a three-element vector that defines the RGB (red-green-blue) coloring of the plot. Finally, create the 3-D plot of z values, and label the X-axis and Y-axis with the x and y values.

```
» mesh(x,y,Z)
» xlabel('X axis')
» ylabel('Y axis')
» zlabel('Z axis')
```

The figure window should show

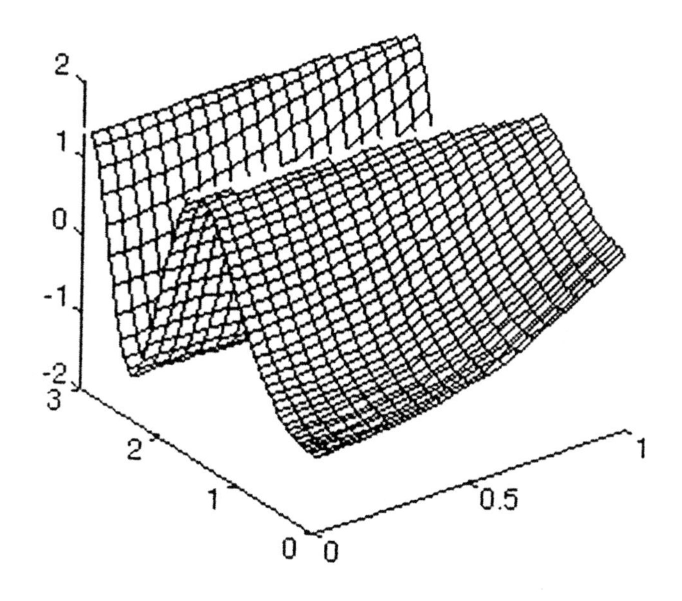

Try It Create a 3-D plot of the function $f(x,y) = \sin(x^{1/2} + y^{1/2})^2$ over the range $x = 0$ to 5 and $y = 0$ to 5 in 0.1 increments.

Subplots

The *subplot* function allows you to place multiple plots in a single window. In this way, you can also print multiple plots on a single piece of paper. The form of the function is *subplot(m,n,p)*, in which the function creates an $m \times n$ matrix of plots, and p (ranging from 1 to $m*n$) is the number of the plot you want to create.

EXAMPLE 4-17

Creating a Set of Subplots in a Single Window

Compare the results of Example 4-16 when using the following four different functions. Show the plot of each function in its own subplot, with the four subplots in a single window.

$$f(x,y) = \sin(x^2 + y^2) + \cos(x^2 + y^2)$$
$$f(x,y) = \sin(x^2 + y^2) - \cos(x^2 + y^2)$$
$$f(x,y) = -\sin(x^2 + y^2) + \cos(x^2 + y^2)$$
$$f(x,y) = -\sin(x^2 + y^2) - \cos(x^2 + y^2)$$

SOLUTION

First, as in Example 4-16, create the vectors **x** and **y** and the matrices **X, Y,** and **Z.**

```
» x=0:0.05:2;
» y=0:0.05:3;
» [X Y]=meshgrid(x,y);
» Z=sin(X.^2+Y.^2)+cos(X.^2+Y.^2);
```

Now use the *subplot* function to create the subplot window with a 2×2 matrix of subplots and activate subplot number 1.

```
» subplot(2,2,1)
```

Now use the *mesh* function to plot the first function, $f(x,y) = \sin(x^2 + y^2) + \cos(x^2 + y^2)$.

```
» mesh(x,y,Z)
```

Now activate subplot number 2 and plot the second function.

```
» subplot(2,2,2)
» Z=sin(X.^2+Y.^2)-cos(X.^2+Y.^2);
» mesh(x,y,Z)
```

Proceed in similar fashion with subplot numbers 3 and 4.

```
» subplot(2,2,3)
» Z=-sin(X.^2+Y.^2)+cos(X.^2+Y.^2);
» mesh(x,y,Z)
» subplot(2,2,4)
» Z=-sin(X.^2+Y.^2)-cos(X.^2+Y.^2);
» mesh(x,y,Z)
```

In the figure window, you would see

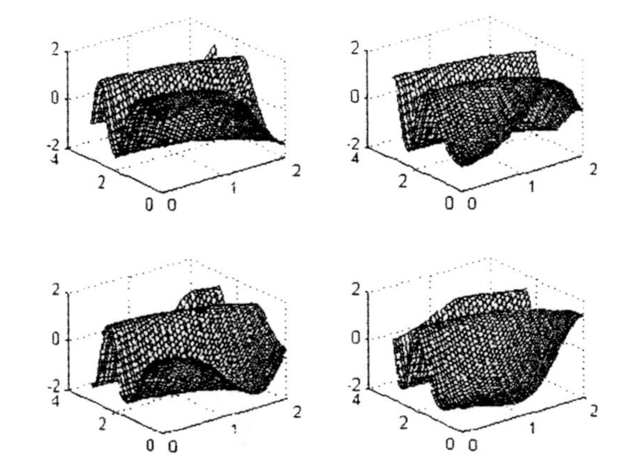

Try It

Use the *subplot* function to create a figure that contains two plots. In the first subplot, display the function $y(x) = \sin(x)/x$ from $x = -20$ to $x = +20$, and in the second subplot, display the function $y(x) = \cos(x)/x$ from $x = -20$ to $x = +20$. (You can ignore any divide-by-zero error messages.)

Animation

The limited space of this module prevents the description of even a small fraction of MATLAB's many advanced functions. However, two of MATLAB's more fun advanced features are its abilities to create animations and movies. As an example, let us create an animation that simulates Brownian motion. We will create 20 randomly placed points and then move them in a random way with (x,y) coordinates between (-1/2,-1/2) and (+1/2,+1/2).

First, clear the workspace and command window, and then set the number of points that will be displayed in the figure window.

```
» clear
» clc
» n=20;
```

Then set a simulated temperature or velocity.

```
» s=0.02;
```

Then create n random points with (x,y) coordinates between -1/2 and +1/2.

```
» x=rand(n,1)-0.5;
» y=rand(n,1)-0.5;
```

Then plot the points in a square with sides at -1 and +1 and use the *set* command to tell MATLAB not to redraw the entire plot when the coordinates of a point are changed, but to restore the background color in the vicinity of the point using an "exclusive or" function.

```
» h=plot(x,y,'.');
» axis([-1 1 -1 1])
» axis square
» set(h,'EraseMode','xor','MarkerSize',18)
```

Lastly, begin the animation using an infinite *while* loop that randomly changes the coordinates of the n points in the plot. Note in the following code that there are no » prompts preceding the last four lines. This is because these lines are a part of the single *while* command. You still press (ENTER) at the end of each line. Pressing (ENTER) after typing *end* will cause the animation to begin, simulating the Brownian motion. You can exit the infinite loop by pressing (CTRL)c.

```
» while 1
x=x+s*randn(n,1);
y=y+s*randn(n,1);
set(h,'XData',x,'YData',y)
end
```

Repeat the preceding animation using 40 points and a simulated temperature of $s = 0.2$.

Application 3: ELECTRIC CIRCUIT MESH ANALYSIS

Electrical Engineering

Electrical engineers must often analyze the circuits they design to determine the power they will consume. Only then can the power supplies be designed. This analysis typically involves the solution of simultaneous equations.

1. Define the Problem

In the circuit shown here, determine the total power delivered to the circuit by the two supplies.

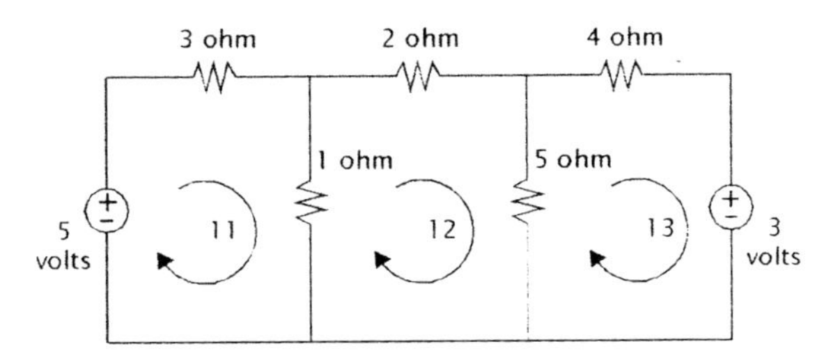

2. Refine the Problem Definition

The power delivered by a power supply is equal to its voltage (V) times the current (I) it supplies to the circuit.

$$P = VI$$

Under certain circuit conditions, a power supply can consume power rather than supply it. When a power supply is providing power to a circuit, its current flows from its positive lead. Therefore, the power provided by the 5-volt power supply in the circuit is

$$P_{5\text{-volt}} = 5I_1$$

Since I_3 is shown flowing from the negative lead of the 3-volt power supply, the power delivered to the circuit by this supply is

$$P_{3\text{-volt}} = -3I_3$$

The total power delivered to the circuit by the two power supplies is

$$P_T = 5I_1 - 3I_3$$

Now, our problem is to determine the two currents, I_1 and I_3.

3. Research Potential Solutions

In order to compute the value of P_T, we need to determine the values of the unknowns, I_1 and I_3. If the circuit has been implemented, we could measure the currents directly using an ammeter. However, we are assuming here that we are analyzing a design that *may* be implemented.

You can determine I_1 and I_3 using a combination of Kirchhoff's voltage law, which states, "The sum of the voltages around a closed path is zero," and Ohm's law, which states, "Voltage equals resistance times current." However, you can use an easier method, a circuit analysis technique known as *mesh analysis,* to write the following equations:

$$(3 + 1)I_1 - (1)I_2 - (0)I_3 = 5 \quad (1)$$

$$-(1)I_1 + (1 + 2 + 5)I_2 - (5)I_3 = 0 \quad (2)$$

$$-(0)I_1 - (5)I_2 + (5 + 4)I_3 = -3 \quad (3)$$

The mesh analysis process allows you to write these equations by simple examination of the circuit. Equation (1) describes the loop labeled I_1, equation (2) describes the loop labeled I_2, and equation (3) describes the loop labeled I_3. Each equation can be described in the following way: the product of the sum of the resistances in the loop and the current in that loop minus the product of any resistance also in an adjacent loop and the current in that adjacent loop is equal to the sum of the voltage rises in the loop.

4. Choose and Implement a Solution

Simplifying the equations provided by mesh analysis, you have

$$(4)I_1 - (1)I_2 - (0)I_3 = 5$$

$$-(1)I_1 + (8)I_2 - (5)I_3 = 0$$

$$-(0)I_1 - (5)I_2 + (9)I_3 = -3$$

Solving this set of simultaneous equations will provide the values of I_1, I_2, and I_3. We need only the values of I_1 and I_3 to compute P_T, the total power we seek. Using MATLAB, we can easily find these values.

You start by noting that you have the following matrix equation:

$$\mathbf{CI} = \mathbf{V}$$

where

$$\mathbf{C} = \begin{matrix} 4 & -1 & 0 \\ -1 & 8 & -5 \\ 0 & -5 & 9 \end{matrix} \qquad \mathbf{I} = \begin{matrix} I_1 \\ I_2 \\ I_3 \end{matrix} \qquad \mathbf{V} = \begin{matrix} 5 \\ 0 \\ -3 \end{matrix}$$

The matrix multiplication, **CI** = **V**, can also be expressed, using the transposes of the matrices, as **I′C′** = **V′**, an easier form for MATLAB to handle. Then

$$\mathbf{I}' = \mathbf{V}'/\mathbf{C}'$$

Now, using MATLAB, create a matrix of the coefficients and matrix of the voltages.

```
» C=[4  -1  0;  -1  8  -5;  0  -5  9];
» V=[5;  0;  -3];
```

Compute the transpose of the current vector.

```
» I=V'/C'

I=
   1.2291   -0.0838   -0.3799
```

Therefore, I_1 = 1.2291 amps, I_2 = -0.0838 amps, and I_3 = -0.3799 amps. Then

```
» PT=5*I(1)-3*I(3)

PT=
   7.2849
```

The total power being delivered to the circuit by the two power supplies is 7.2849 watts.

5. Test the Solution

To verify this solution mathematically, replace the variables, I_1, I_2, and I_3 in the original set of simultaneous equations with their computed values.

$$(4)(1.2291) - (1)(-0.0838) - (0)(-0.3799) = 4.9164 + 0.0838 = 5.0002$$

$$-(1)(1.2291) + (8)(-0.0838) - (5)(-0.3799) = -1.2291 - 0.6704 + 1.8995 = 0$$

$$-(0)(1.2291) - (5)(-0.0838) + (9)(-0.3799) = 0.4190 - 3.4191 = -3.0001$$

With some small rounding errors, the computed current values do solve the original mesh equations. The solution could also be verified by constructing the circuit and then directly measuring the current, using an ammeter, from each power supply. Once the current is known, you can calculate the power from

$$P_T = 5I_1 - 3I_3$$

What If What if the polarities of the two power supplies were reversed? Would they still deliver 7.2849 watts to the circuit?

SUMMARY

This chapter covered matrix functions, data analysis, and plots. The matrix function subjects covered in the first section of this chapter included solving nonlinear equations, solving sets of linear equations, finding roots of polynomials, linear and polynomial regression, and numerical differentiation and integration. Section 4-2 described data analysis subjects, such as random number generation and finding minima, maxima, means, and standard deviations. The section concluded with an explanation of histogram creation. Section 4-3 briefly described MATLAB's plotting capabilities, including the creation of simple bar graphs and polar, contour, and 3-D mesh plots.

Key Words

function functions
interpolation
least squares fit
linear regression
matrix function
polynomial regression

MATLAB Functions

bar(y)	Creates a bar graph of the elements in vector **y**
ceil(x)	Returns the integer nearest to and greater than *x*
colormap(c)	Sets the color scheme for plots; **c** is a three-element RGB vector
conj(c)	Returns the conjugate of complex number *c*
contour(X)	Creates a contour plot of the elements in matrix **X**
diff(f)	Returns the differences between consecutive elements in **f**
fix(x)	Rounds **x** toward zero
floor(x)	Returns the integer nearest to and less than *x*
fmin('fname',L1,L2)	Finds the local minimum of *fname* between the limits *L1* and *L2*
fplot('fname',L,n)	Plots *n* points of *fname* between the two limits in **L**
fzero('fname',X0)	Finds the zero of *fname* near the value *X0*
hist(x)	Creates a histogram of the elements in vector **x**
imag(c)	Returns the imaginary part of complex number *c*
max(x)	Returns the maximum value in vector **x**
mean(x)	Returns the means of the elements in vector **x**
median(x)	Returns the median of the elements in vector **x**
mesh(Z)	Creates a 3-D plot using the values of the elements of the matrix **Z** as the Z-axis values
meshgrid(x,y)	Used to help create plots of a function of two variables
min(x)	Returns the minimum value in vector **x**
polar(theta,r)	Creates a polar plot of the angles in vector **theta** versus the radii in vector **r**
polyfit(x,y,n)	Returns the coefficients of a polynomial of degree *n* that best fits the data points (x_i, y_i) in a least squares sense
prod(x)	Returns the product of the elements in vector **x**

quad('fname',a,b)	Returns the result of integrating the function '*function*' over the interval *a* to *b*
rand(n)	Returns an $n \times n$ matrix whose elements are uniformly distributed random numbers in the interval (0,1)
rand(m,n)	Returns an $m \times n$ matrix whose elements are uniformly distributed random numbers in the interval (0,1)
randn(n)	Returns an $n \times n$ matrix whose elements are normally distributed random numbers in the interval (0,1)
real(c)	Returns the real part of complex number *c*
roots(C)	Returns a column vector whose elements are the roots of the polynomial function whose coefficients are in vector **C**
round(x)	Rounds **x** to the nearest integer
sort(x)	Returns the vector **x** with its elements sorted into ascending order
std(x)	Returns the standard deviation of the elements in vector **x**
subplot(m,n,p)	Creates a window of $m \times n$ plots with plot number *p* as the active plot
sum(x)	Returns the sum of the elements in vector **x**

Problems

1. Define the function $f(x) = \cos(x)/x$ in an M-file. Plot $f(x)$, and determine the minimum and zero of $f(x)$ between $x = 0$ and $x = 10$.
2. Write a program that solves quadratic equations. Have the user of the program enter *a*, *b*, and *c*, the coefficients of $y(x) = ax^2 + bx + c$.
3. Solve the following set of simultaneous equations:

$$\begin{aligned} 2x_1 - 2x_2 + x_3 &= 2.5 \\ -2x_1 + 3x_2 + 2x_3 &= -5.1 \\ -x_1 + 2x_2 + x_3 &= 0.2 \end{aligned}$$

4. Determine the roots of the polynomial function $f(x) = x^4 - 3x^2 - 1$. Then, create an M-file definition of this function and check the roots to confirm that they are correct.
5. Write a program that plots least squares fit curves through data, entered by the user of the program, that represent the daily measured pressure, in psi, in a tank. The user enters as many daily measurements as desired. An entry of zero indicates that the preceding measurement was the final measurement. Test the program with the following data: 11, 13, 17, 16, 20, 20, 23, 25, 25, 31, 30, 34, 36, 41, 41, 44, 44, 47, 53, 54, 55.
6. Determine the derivatives of the following functions at the point $x=0$:
   a. $y(x) = 3x^2 - 5x + 1$
   b. $y(x) = -3x^4 + 3x$
   c. $y(x) = 5e^{3-4x}$

7. Determine the derivatives of the following expressions at the point $x = \pi/8$:
   a. $\sin(2x)^2$
   b. $-\cos(4x)^{-2}$
   c. $5e^{4-3x}$
   d. $\cos(x^{-2})^2 - \sin(x^{-2})^2$
8. Write a function that accepts three input arguments: the vectors **x** and **y** and an index **i.** It returns the derivative of **y** with respect to **x** calculated at $\mathbf{x}_i$. The function returns that derivative with an accuracy of 0.001. (Hint: This will require a loop in which the calculated result from one iteration of the loop is compared to the result from the previous loop.)
9. Determine the integral of the following functions over the interval $x = -2$ to $x = +2$:
   a. $y(x) = 4x^2 - 5x + 1$
   b. $y(x) = -3x^4 + 2x$
   c. $y(x) = 5e^{4-3x}$
10. Determine the integral of the following functions over the interval $x = -\pi/2$ to $x = +\pi/2$:
    a. $y(x) = \sin(2x)^2$
    b. $y(x) = -\cos(4x)^{-2}$
    c. $y(x) = 5e^{4-3x}$
11. Create a 100-element vector containing random real numbers that range uniformly from 0 to 600 meters with a normal distribution. Then plot a histogram that shows how many of these random numbers lie between 0 and 100, 100 and 200, ..., and 500 and 600 meters. Determine the minimum and maximum of the elements in the vector. Determine the mean and standard deviation of the elements in the vector.
12. Create a 50-element vector containing random real numbers that range from -10 to 10 with a normal distribution. Sort the vector elements, and then plot a bar graph of the elements.
13. Create a single polar plot of the two functions $r(\theta) = 1.1^{-0.25\theta}$ and $g(\theta) = 1.1^{0.25\theta}$.
14. Create a single polar plot of the two functions $f(\theta) = \sinh(\theta^{-1/2})$ and $g(\theta) = -\cosh(\theta^{-1/2}) + 1$, $\theta = 1.360$ degrees, in 1-degree increments.
15. Determine the low and high points of the surface defined by $f(x,y) = \sin(x^2 + y^2) - \cos(x^2 + y^2)$ for $x = 0$ to 4 and $y = 0$ to 4 in increments of 0.1. Create contour and mesh plots of the function.

Use the five-step problem-solving process while working Problems 16 through 20.

16. Assume a manufacturing process in which the overhead cost is \$50,000. It costs \$18.50 to manufacture each part. Graphically determine the economic breakeven point if the parts are sold for \$30.00 each.

17. The following represents daily temperature highs, in degrees Celsius, each day for ten days:

day =	1	2	3	4	5	6	7
temperature =	26.1	27.0	28.2	29.0	29.8	30.6	31.1

day =	8	9	10
temperature =	31.3	31.0	30.5

Perform a linear regression and a third degree polynomial regression on the data points. Decide which regression might best predict the high temperature on day 11 and what that high temperature might be.

18. Assume the speed of an object is defined by

$$S(t) = -t^2 + 100t$$

$S(t)$ is in meters per second, and t is in seconds.

a. Determine the time at which the object reaches its highest speed.

b. Determine the highest speed reached.

c. Determine the distance the object travels between $t = 0$ and $t = 100$ seconds.

19. Graph the function $y(t) = 490 - (a/2)(t - 10)^2$, where a equals 9.8 m/sec^2, the acceleration due to gravity; t is in seconds; and $y(t)$ is the altitude in meters. Determine the time t' at which the altitude is zero again, after starting at zero.

20. Assume you are a civil engineer who must make a rough estimate of how many cubic yards of earth must be removed in order to level a plot of land to 5 feet above sea level. The following data represents elevations measured, in feet, every yard in each direction on the plot, the size of which is 7 by 7 yards.

```
5 5 5 5 5 5 5 5
5 6 6 6 6 6 6 5
5 6 7 7 7 7 6 5
6 7 7 8 8 7 7 6
6 7 7 8 8 7 7 6
5 5 5 5 5 5 5 5
5 6 6 6 6 6 6 5
5 6 7 7 7 7 6 5
```

Store this 8×8 matrix in an ASCII character file called MATRIX.DAT. Read the matrix, and create a contour plot of its contents. Make a rough estimate of the amount of earth that must be moved through visual examination of the contour plot.

21. Create a figure that contains four plots. Display the function $y(x) = \sin(x)/x$ in all four plots. However, in the first plot, display the function from $x = -20$ to $x = +20$. In the second subplot, display the function from $x = -10$ to $x = +10$. In the third subplot, display the function from $x = -5$ to $x = +5$. And in the fourth subplot, display the function from $x = -2$ to $x = +2$. Make sure all plots show smooth lines. Include girds on the plots. (You can ignore any divide-by-zero error messages.)

Index

O

LA DWP 213 367 4211
SC Gas Co 213 244 1200